Anonymous

Biennial Report of the Trustees, Superintendent and Treasurer of the Illinois Southern Hospital for the Insane

Vol. 4

Anonymous

Biennial Report of the Trustees, Superintendent and Treasurer of the Illinois Southern Hospital for the Insane
Vol. 4

ISBN/EAN: 9783337372972

Printed in Europe, USA, Canada, Australia, Japan

Cover: Foto ©berggeist007 / pixelio.de

More available books at **www.hansebooks.com**

FOURTH BIENNIAL REPORT

OF THE

TRUSTEES, SUPERINTENDENT AND TREASURER

OF THE

ILLINOIS

SOUTHERN HOSPITAL FOR THE INSANE,

AT ANNA.

October 1, 1880.

———

SPRINGFIELD:
H. W. ROKKER, STATE PRINTER AND BINDER.
1881.

OFFICERS OF THE HOSPITAL.

REPORT OF TRUSTEES.

Hon. Shelby M. Cullom, *Governor of Illinois:*

Sir:—We herewith submit the fourth biennial report for the Illinois Southern Hospital for the Insane.

At the date of our last report, in 1878, an execution had been issued on the judgment obtained against the former treasurer of this hospital and his sureties. The judgment was for two thousand seven hundred and eighty-one dollars and seventy cents, for principal and interest, as then stated, and was rendered on the 18th of April, 1878.

On the 18th of November, 1878, the judgment, with interest and nineteen dollars and fifty cents costs of suit added, amounted to two thousand eight hundred and ninety-eight dollars and fifty-five cents. On that day two thousand one hundred and seventy-three dollars and ninety-two cents, three-fourths of the amount then due, was paid to the sheriff of Williamson county, out of which he retained the sum of nineteen dollars and fifty cents, costs of suit, and on the 30th of November paid over to the president of the board the balance in his hands, two thousand one hundred and fifty-four dollars and forty-two cents, which amount the president paid over to the treasurer of the hospital, and took his receipt therefor.

On the 30th of November, 1878, there remained due on the judgment a balance amounting, with interest, to the sum of seven hundred and twenty-six dollars and two cents. Some real estate had been levied upon, and that was the day of sale. The president of the board attended the sale and bid off, in the name of the trustees, for the use of the state, various tracts of land which were redeemed last February, the amount with interest due at the time of redemption being eight hundred and fifteen dollars and fifty-five cents, for which a draft was sent to the president for eight hundred and fourteen dollars and eighty cents, a charge of seventy-five cents being made for the draft. A receipt was taken from the treasurer of the hospital for the amount of the draft, which was duly paid, and that closed the matter, the judgment being fully liquidated. The balance due from the former treasurer, as stated in our last report, was two thousand six hundred and fifty-three dollars and thirty cents, for which amount, and one hundred and twenty-eight dollars and forty cents interest, judgment was given in favor of the hospital on the 18th of April, 1878. The amount collected on it, with interest, was two thousand nine hundred and sixty-nine dollars and twenty-two cents, that sum being the total of the redemption money, eight hundred and fourteen dollars and eighty cents for land sold, and two thousand one hundred and fifty-four dollars and forty-two cents collected and paid over by the sheriff of Williamson county.

The new barn, in course of construction at the date of our last report, was finished as we could from time to time apply money from the fund for improvements and repairs to that purpose.

The old barn, for the removal of which the last general assembly made an appropriation of one thousand dollars, was removed to a place on line with and near the new barn. It was put on sloping ground, on a strong rock foundation, and refitted as a bank barn, for the amount appropriated for that purpose, and contains more storage room for feed and other uses than the new one. The basement was fitted up for cattle and young stock, with a vegetable cellar at the south end.

Both the new and old barns were so heated as to afford ample natural drainage in such a direction as not to contaminate the water supply of the hospital. All the water needed at the barns is conveyed to them through pipes.

Shortly after the date of our last report a duplex Worthington pump was erected near the boiler dam on the east side of the hospital, to be used in case of fire. Hose was purchased and arrangements so perfected that, in case of fire, a stream of water can be turned on in from two to three minutes' time, when necessary, at so many different places in the hospital and at the barns, that there is now very little, if any, cause for apprehension on account of fire. A fire corps of male attendants has been organized, which is drilled occasionally so as to familiarize them with their duties in case of necessity. It required the whole amount—eighteen hundred dollars—appropriated for that purpose. The machinery needed was procured before the advance in the price of iron.

The water supply of this hospital was a source of uneasiness and apprehension to us after examining it soon after entering on our duties, and attention was called to it by the retiring superintendent and his successor, in the third biennial report. The supply was mainly surface water, collected in a dam into which the rain water flowed from a surface of about thirty acres. and partly from a spring which was made to flow into a large cistern near a pump-house at the dam. The quantity of water required for the daily supply of the hospital is upwards of thirty thousand gallons. Two years ago there was not much rain during the fall season, and on the 28th of October, 1878, the dam failed entirely, and the spring flowing into the cistern yielded only about five thousand gallons of water in 24 hours. Upon being notified of the fact, the president of the board came to the hospital at once, and, with the resident trustee, made another inspection of the grounds and of a spring called the "big spring," on the land belonging to the hospital, about one-third of a mile south-east of the building, and the flow of water was measured as accurately as the means at hand, and which could be devised, permitted. The emergency was great, and the difficulty had to be surmounted without delay. Water had to be procured at once, or the hospital practically closed. Fortunately the erection of the fire pump was so nearly completed that ·by working day and night the pump was made available on the 5th of November, 1878, and with it the emergency was tided over by pumping water from the boiler dam for the time being.

After testing the flow of the "big spring," the conclusion was that it would yield a sufficient supply of water, and a contract was at once made with John Davis & Co., of Chicago, to furnish a pump of sufficient capacity and about sixteen hundred feet of four inch iron pipe, for the sum of fourteen hundred and twenty-nine dollars. Work was commenced at once. A small pump house was built, the pump erected, water pipe laid and temporarily connected with the pipe under the center building of the hospital and extending thence to the elevator tank, so that water could be supplied from the spring within about seven weeks from the time the water in the dam failed. A small boiler, no longer used in the old pump house, was taken out, repaired and refitted temporarily to supply steam for the new pump.

To the energetic management of Doctor Wardner, the superintendent, and the willing assistance of Mr. John Davis, who was on the ground, part of the time superintending his workmen day and night, we feel under obligations for being able to have the hospital supplied with water so soon after the almost entire failure of the supply provided for when it was built.

The amount required to meet the emergency, so that water could be supplied, fourteen hundred and twenty-nine dollars, was paid out of the ordinary expense fund of the hospital, which increased the *per capita* expense to that amount. The matter could not be deferred until the general assembly would meet and a special appropriation be made. We trust our action will be approved.

The original arrangements for supplying water were left intact, and the machinery is kept in order.

The spring has now been tested for nearly two years, and we feel no more solicitude about the supply of water.

The appropriation of twenty-five hundred dollars for water supply, made by the last general assembly, so far as it has been expended, has been used to make a direct connexion between the new water pump and the tank from which the buildings are supplied, and to purchase, lay and extend pipes needed to perfect the system of water supply. It also became necessary to replace the lead pipes in the hospital with iron pipes, in consequence of rats gnawing through the lead pipes so much as to be a constant source of annoyance and cause damage to the buildings by leakage. Part of the appropriation was applied to that work.

A new kitchen, as completely furnished as it could be, has been built for the amount appropriated for that purpose by the last general assembly. It is a substantial building, conveniently arranged, supplied with hot and cold water pipes, the necessary utensils, and furniture, and is in every respect much better than the old one. The space formerly occupied by the old one is being utilized for other purposes.

Within the last two years the grounds have been very much improved, nearly all the driveways and walks have been made and covered with gravel, many native trees were planted and much

8

grading done. Flower borders were made on the sides of the driveways and walks through the lawn in front of the hospital, and filled with a variety of plants, making a cheering and pleasing appearance. The unsightly high board fence which enclosed a court yard, has been taken down, and the grounds are assuming a more attractive appearance.

The original plan for ornamenting the grounds embraced two fountains, one opposite each wing. We have concluded to have only one erected, opposite the center building, at such a distance from it as to be visible from all parts of the hospital. Some work has been done on it.

The sewer has been extended to such a distance that the offensive effluvia from the outlet thereof no longer reaches the hospital as it did formerly, when the outlet was just below the boiler dam.

The concrete paving in the wash-room became so worn and rough that it remained in a wet state constantly, and looked more like a muddy piece of turnpike than the floor of a wash-room in which human beings had to labor. It has been replaced with an asphalt brick pavement, a lasting material, making a smooth floor, which has greatly improved the room.

Some rooms in the most convenient place in the basement have been fitted up for store-rooms and paved with asphalt brick. They are excellent rooms for the purpose for which they are used.

We have endeavored to make contracts for furnishing staple supplies by advertising for proposals, and contracts have been made for furnishing meat and fuel. They were awarded to the lowest bidders, and bonds were required and given for the fulfillment of their contracts. For coffee, sugar and tea we could get no satisfactory bids for any length of time, dealers saying that prices fluctuated too much. For furnishing flour we advertised for proposals several times within the last year, but received no bids which we thought it advantageous to accept, and directed the superintendent to purchase on the market wherever he could buy the best article at the lowest price. For supplies not furnished on contract, bids are practically received when purchases are made, and, as a general thing, supplies have been bought cheaper than bidders proposed to furnish them.

On the 30th of September, 1880, there were five hundred and one patients in the hospital. Since the introduction of spring water and the extension of the sewer, the health of the inmates has improved, and the death-rate has greatly decreased. There has been a marked increase in the number of recoveries and improvement of patients in the last two years, due in a large measure, we think, to the able management and direction of Doctor Wardner, and his assistants, Doctors Hester and Stocking, as well as the improved sanitary condition of the hospital. For the sake of the unfortunate inmates, it is very gratifying to us to feel justified in making this statement.

In the accompanying report of the superintendent, annexed tables, and cash statement of the treasurer, will be found the information and details required to be furnished by sec. 28 of the "Act to regulate the state charitable institutions," approved April 15, 1875.

We invite thoughtful and special attention to the improvements recommended by the superintendent, and the requests he makes. Having carefully considered them, we give them our earnest endorsement.

On the 30th of September, 1880, there were six convict patients in this hospital, two of them under sentence of confinement for life. Before the completion of this report, two more were brought here from the Joliet penitentiary, also under sentence for life. In our judgment, the mixing of convict patients with those guiltless of any offense has a very detrimental effect. We would earnestly urge that some other means be devised for the separate care of convict patients. Considerable extra expense would have to be incurred to give them more precautionary care than is given to those not under sentence, and they are liable to make their escape. Hospitals cannot be converted into prisons without extra cost.

We have maturely considered the appropriations asked for, and the purposes for which they are asked, and would earnestly recommend and urge that they be granted by the general assembly. In view of the possible, if not probable, advance in price of at least some supplies, we consider the amount asked for ordinary expenses as barely sufficient.

The engine now used in the hospital is not powerful enough to run all the machinery required, and one of greater power is an imperative and pressing necessity. About double the present power is needed.

The amount asked for improvements and repairs will be barely sufficient. The flooring in the basement is now fast wearing out in places, and as it needs repairing, the intention is to pave with asphalt brick. In making improvements and repairs, we have endeavored to use as lasting and indestructible material as possible, with a view to economy and lessening expenses for repairs in the future. We will not undertake to enumerate all the repairs needed. In such an institution, containing many destructive persons, they are necessarily numerous, independent of the repairs needed on account of ordinary wear and tear.

The appropriation asked for improvement of grounds will be needed to improve them according to the original plan, and give them a beautiful and attractive appearance. Much of the labor has been and will be done by patients, but much grading remains to be done and teams have to be hired, and other expenses incurred.

A small appropriation is asked for fencing, with a view to constructing a more sightly fence, about half a mile in length, along the state land on the east side of the road from the hospital to Anna, where there is now only an ordinary worn rail fence. A cedar post and picket fence can be constructed for the amount asked, and would improve greatly the appearance of the property of the state.

The hospital has been repeatedly struck by lightning, and we call attention to what the superintendent says on that subject, and the appropriation to put up lightning-rods.

We also call attention to the plans and specifications for a refrigerating-house, and reservoir and filterer for water—to what the superintendent says on those subjects, and to the special appropriations asked for those purposes. The plans were made by a practical architect and civil engineer, and we submit them to your consideration after having fully considered the necessity for both improvements.

The special appropriation asked for the purpose of improving the ventilation of the north wing, and for machinery for the carpenter-shop, also have our endorsement and approval. The defective ventilation of that wing has been made the subject of comment in previous reports. We will only add, that we hope we may be enabled to have it improved, if not perfected.

In administering the affairs of the hospital, we have endeavored to use the utmost economy consistent with the welfare of the unfortunate inmates and the interests of the state. No unnecessary expenditures were made that we are aware of. The fire-pump, a large quantity of iron water-pipe, and the pump and machinery at the big spring furnishing the new supply of water, were procured while the price of iron was low.

An inventory of the property under our control and care has been made, and also a careful appraisement of its value. A statement thereof is annexed. The valuation annexed to our last biennial report was too low,

To Doctor Wardner, the medical superintendent, and the assistants and officers under his direction, commendation is due for the faithful, diligent and efficient performance of the duties devolving on them.

Respectfully submitted, together with the report of the superintendent and annexed statements.

John E. Detrich, *President.*
E. H. Finch,
W. P. Bruner,

October 21, 1880.   *Trustees.*

REPORT OF THE SUPERINTENDENT.

To the Trustees:

GENTLEMEN—As required by the statute regulating charitable institutions, I present you the following report concerning operations in this hospital for the two years ending September 30, 1880:

At the date of last report there were remaining in the hospital two hundred and fifty male and two hundred and eighty female patients.

During the two years just past, there have been admitted one hundred and eighty-one males and one hundred and forty-seven females, making a total of seven hundred and eighty-six under treatment. Of this number one hundred and thirty-six have been discharged recovered, twenty-six much improved, twenty-seven improved and forty-two stationary.

The number of deaths the first year was thirty-three, the second year twenty-one.

The per cent. of recoveries on admissions is forty one and forty-six hundredths, and on the whole number under treatment seventeen and three-tenths.

The following tables will set forth more concisely the facts relating to the inmates:

TABLE I.

Movement of Population.

General Results.	1878-9			1879-80		
	M.	F.	Total	M.	F.	Total
Number remaining October 1, 1878	250	208	458	271	228	499
Admitted since	107	83	190	74	64	138
Whole number treated	357	291	648	345	292	637
Discharged:						
Recovered	39	34	73	35	28	63
Much improved	6	7	13	8	5	13
Improved	5	8	13	7	7	14
Not improved	15	2	17	20	5	25
Died	21	12	33	13	8	21
Total discharged and died	86	63	149	83	53	136
Remaining September 30 of each year	271	228	499	262	239	501
Average present in each year	249.8	215.7	465.58	261.3	236.8	498.23

TABLE II.

Age, Sex and Civil Condition of those Admitted.

Age, etc.	September 1878-80. Males.				September, 1878-80. Females.			
	Single.	Married.	Widowed.	Total.	Single.	Married.	Widowed.	Total.
Under 10 years	1			1	1			1
From 10 to 15 years					2			2
From 15 to 20 "	16			16	9	3		12
From 20 to 25 "	28	1		29	10	18		28
From 25 to 30 "	24	8	1	33	6	11		17
From 30 to 35 "	6	10	1	17	4	16	2	22
From 35 to 40 "	5	17	1	23	2	20	3	25
From 40 to 45 "	3	19	2	24		6	9	15
From 45 to 50 "	1	14	2	17	2	6	2	10
From 50 to 55 "		5		5	1	3	1	5
From 55 to 60 "		7	1	8		3	3	6
From 60 to 65 "		3	1	4		3		3
From 65 to 70 "	1		1	2		1		1
Age unknown	1		1	2				
Totals	86	84	11	181	37	90	20	147

TABLE III.

Form of Mental Disorder in those Admitted.

Mental Disorder	1878-9			1879-80		
	M.	F.	Total	M.	F.	Total
Mania, acute	39	34	73	31	12	43
Mania, chronic	30	7	37	15	17	32
Mania, puerperal		9	9		4	4
Mania, epileptic	7	5	12	10	3	13
Mania, periodic		7	7		10	10
Melancholia, acute	13	12	25	12	7	19
Melancholia, chronic	7	3	10	3	1	4
Dementia, acute		2	2		2	2
Dementia, chronic	8	4	12	3	6	9
Melancholia, periodic					2	2
Moral insanity	1		1			
Idiocy	2		2			
Totals	107	83	190	74	64	138

TABLE IV.

Occupation of those Admitted.

Occupation.	1878-79.			1879 80.		
	M.	F.	Total	M.	F.	Total
Architect	1		1			
Agent				1		1
Barber	1		1	1		1
Blacksmith	3		3	1		1
Bootblack	1		1			
Clerk				2		2
Capitalist	1		1			
Carriage-trimmer	1		1			
Domestic		3	3		1	1
Dressmaker		1	1			
Engineer	2		2			
Farmer	66		66	44		44
Fireman				1		1
Harnessmaker	1		1			
Housekeeper		76	76		60	60
House-work		1	1			
Laborer	14		14	10		10
No occupation	3	1	4	2	3	5
Merchant	3		3	1		1
Machinist				1		1
Miner				2		2
Miller	1		1			
Marble dealer	1		1			
Printer	1	1	2			
Physician	1		1	2		2
Painter				1		1
Saloon-keeper	1		1	1		1
Shoemaker	2		2	1		1
Stonemason	1		1			
Student	1		1	1		1
Tanner	1		1			
Teacher				1		1
Wagonmaker				1		1
Totals	107	83	190	74	64	138

TABLE V.

Counties from which Patients were Admitted.

County.	1878-79.			1879-80.		
	M.	F.	Total	M.	F.	Total
Alexander	4	3	7	1		1
Bond	1	4	5	3	2	5
Clinton	2	5	7	1	1	2
Crawford		1	1			
Clay	1	3	4	3		3
Clark	3	1	4	1	4	5
Coles	4	3	7	4	6	10
Cook				1		1
Cumberland				1	1	2
Champaign	3	3	6	2	3	5
Christian				1		1
Douglas	1		1		2	2
Edwards	1		1	1		1
Effingham	2	3	5	1		1
Fayette	1	1	2	3	1	4
Franklin	4		4	2	2	4
Gallatin	4		4	2	1	3

TABLE V—*Continued.*

Occupation.	1878-79.			1879-80.		
	M.	F.	Total	M.	F.	Total
Hardin				1		1
Hamilton	2		2	2	1	3
Jefferson	1	1	2			
Jasper	3		3	3	2	5
Johnson	2	2	4			
Jackson	3	4	7	2	3	5.
Lawrence	5	1	6			
Monroe	3	2	5	2	4	6
Massac	2	1	3	1	1	2
Marion	3	2	5	5		5
Madison	4	6	10	3	6	9
Moultrie		3	3	2		2
Pulaski	2		2	1	1	2
Perry	2	1	3		2	2
Richland	1		1		2	2
Randolph	6	7	13	4		4
St. Clair	4	6	10	6	2	8
Saline	1	1	2	1		1
Shelby	7	4	11	3	2	5
Union	2	2	4	3	2	5
Vermilion	8	5	13	2	7	9
Williamson	4	1	5		1	1
Wayne	3		3	3	1	4
White	2	2	4	2	1	3
Washington	3		3	2	1	3
Wabash	2	4	6		1	1
Warren	1		1			
Totals	107	83	190	74	64	138

TABLE VI.

Deaths and Causes.

Causes.	1878-9.			1879-80.		
	M.	F.	Total	M.	F.	Total
Angina pectoris					1	1
Apoplexy		1	1			
Cardiac embolism	2		2			
Congestion brain and spinal cord	1		1			
Congestion of lungs				1		1
Compression of brain from injury	1		1			
Dysentery acute		1	1			
Epilepsy	2		2	3		3
Erysipelas	1	1	2			
Exhaustion acute mania	2	1	3	1		1
Exhaustion chronic mania	1		1		1	1
Exhaustion epileptic mania		1	1		1	1
Exhaustion acute melancholia	1		1			
Exhaustion chronic dementia				1		1
Fatty degeneration of heart	1		1			
General paralysis				1		1
Inanition	1		1	2		2
Organic brain disease	1		1			
Phthisis pulmonalis	5	5	10	4	3	7
Stricture of bowels				1		1
Suicide	1	1	2			
Syphilis				1		1
Traumatic meningitis	1		1			
Uremia (occlusion of urethra)		1	1			
Totals	21	12	33	13	8	21

TABLE VII.

Nativity of those Admitted.

Nativity.	1878-9.			1879-80.		
	M.	F.	Total	M.	F.	Total
United States	92	68	160	60	55	115
Germany	10	11	21	9	6	15
Ireland	3	2	5	4	2	6
England	1	1	2		1	1
Switzerland		1	1			
France				1		1
Canada	1		1			
Totals	107	83	190	74	64	138

TABLE VIII.

Number of Attack in those Recovered.

Number of Attack.	Male.	Female.	Total.
First	51	43	94
Second	16	13	29
Third	4	3	7
Fourth		1	1
Fifth	1		1
Ninth	1	1	2
Unknown	1	1	2
Totals	74	62	136

TABLE IX.

Operations of the Hospital, year by year, from the opening, December 15, 1873, to September 30, 1880.

Years.	Admitted.			Discharged.															Remaining Sept. 30, in each Year.			Average No. Resident in each Year.		
				Recovered.			Improved.			Stationary.			Died.			Not Insane.								
	M.	F.	T.	M.	F.	T.	M.	F.	T.	M.	F.	T.	M.	F.	T.	M.	F.	T.	M.	F.	T.	M.	F.	T.
1873–4	96	62	158	6	4	10	...	2	2	2	3	5	5	3	8	...	...	...	83	50	133	94.5	54.1	148.6
1874–5	67	36	103	9	5	14	15	6	21	21	12	33	6	1	7	...	...	...	99	59	158	129	75	204
1875–6	88	58	146	14	15	29	11	5	9	6	4	6	8	11	19	...	...	...	159	82	241	161.5	84.1	245.6
1876–7	56	36	92	21	12	32	11	2	13	4	2	17	7	10	17	...	1	1	173	90	263	226.3	167.4	393.7
1877–8	136	152	308	31	21	52	11	4	18	16	1	17	15	11	26	...	...	...	553	265	458	219.8	215.7	465.5
1878–9	167	83	190	38	34	73	15	15	26	15	2	25	21	12	33	...	...	...	271	228	499	261.3	236.8	498.2
1879–80	74	61	138	35	28	68	15	12	27	20	5		13	8	21	...	...	1	262	239	501			
Totals...	644	491	1,135	154	119	273	70	49	119	80	29	109	75	56	131	...	1	1						

It will be observed in the foregoing tables that while the number of patients has increased, the number of deaths has been thirty-six and three-tenths per cent. less the last year than during the year preceding. This I believe to be largely due to improved sanitation, which will receive attention further on under appropriate heads.

Of the number of cases now under treatment fifty-one are epileptic, four hundred and fifty chronic and probably incurable, fifty-one are acute, fifty-one probably curable, eighty-four are destructive, and sixty-one require personal care on account of their demented state and uncleanly habits.

The eighty-four destructive patients include those who are homicidal and suicidal. It has been the constant aim in the treatment of these cases to limit the use of the means of restraint to the minimum. Yet, for the safety of the patients themselves and others, it has been found necessary to use it in a mild form, which leaves the patient comparatively free to exercise as he may desire. The average number of patients who have thus been restrained is about two per cent. of the average number treated.

In the treatment and care of the insane, kindness has been insisted upon as an essential in the conduct of the attendants, and one of the chief aims has been to allow as much liberty as possible within the limits of safety. It has also been an object to keep the mind diverted by amusements and congenial employment, without taxing the physical energies or requiring more than would be necessary as healthful exercise.

About twenty-five, selected daily from the more feeble, have been driven in an easy carriage two or three miles every pleasant day for the past year and a half. This has produced much evident benefit to these invalid cases. The stronger among the women have been arranged in details to assist in the work about the domestic department by turns; half day at a time. It is entirely optional with the patients to assist or not. They are generally very glad of the opportunity. Those who have the taste and desire to make fancy work are furnished with material for that purpose. This branch of employment is now self-sustaining, the sale of articles made affords means to supply material for the work.

There has been an average of about twenty-five male patients who have worked on the farm and grounds a good portion of the time. It is particularly noticeable that the men who help outside in the work are greatly improved in health and physical appearance.

All patients not otherwise engaged are taken out in companies, once or twice daily for a walk of an hour or more, so that all are either employed or have out-door exercise,

The airing courts, enclosed by high, tight board fences, fell into disuse over a year ago, and have now been removed as useless, unsightly and obnoxious to the patients.

I have had much annoyance in caring for the criminal insane. The law regarding them is very defective. It makes no provision for notifying the superintendent, consequently they are thrust in unannounced whether there is room or not.

2—

The law is also silent upon the subject of the expense of returning those who recover, to the prison; and makes no provision for expenses of clothing. It also makes no provision for the treatment of those who are still insane when the term of their sentence expires. It says they "shall be treated as other patients." Other patients would be retained as long as necessary; but the attorney-general has decided that the hospital has no legal right to keep convict insane beyond the expiration of sentence.

The prison commissioners have refused to furnish the usual allowance given to convicts at their discharge; the counties cannot be held for the expense of their return, and they naturally try to avoid any responsibility on their account. So it may often happen that the superintendent must choose between turning a dangerous lunatic loose upon community, and retention in hospital without due authority, and at his own peril.

Let the law be amended on this point, or what is far better, and more in keeping with our ideas of social relation, let there be at once established a hospital in connexion with one of the prisons, in which all convict insane and criminal insane can be kept by themselves. It is a shame to thrust the insane from the criminal classes upon the society of the law-abiding citizens who are so unfortunate as to suffer mental alienation.

Patients being brought to the hospital are often deceived by their friends as to the object of their coming. This course often makes a lasting impression upon their minds, and they think of, and grieve over the imaginary wrong even after convalescence. It were far better to deal with them fairly and honestly, that they may learn to trust their friends, and those who have the care of them. Representations or promises should never be made to them which are untrue or cannot be fulfilled; otherwise they suffer, in addition to the delusions of their diseased mental state, from a keen sense of injury unintentionally inflicted by those upon whom they must rely in their sad condition.

RELIGIOUS EXERCISES, AMUSEMENTS, ETC.

Patients have been assembled, in the pleasant room provided for that purpose, as often as seemed practicable.

They have attended religious services regularly every Sabbath afternoon, at which ministers of different denominations have officiated.

During the week entertainments of various kinds, calculated to divert their minds and make hospital life pleasant and more salutary, have been presented to them.

The regular weekly dance they never seem to tire of, and many who do not participate in the exercises enjoy the music and the relief afforded from the monotony of their wards. The other entertainments, given every two or three weeks during the amusement season, have consisted of plays, minstrels, readings, vocal and instrumental music, tableaux exhibitions with the sciopticon, masquerade balls, etc. Nearly all of these amusements have been furnished by the employés of the hospital, and at very little expense.

We have also usually been able to maintain among the employés an orchestra of five or six instruments, which has furnished excellent music for dances and other entertainments.

Perhaps nothing has been enjoyed more by the patients than the homelike and generous Christmas and Thanksgiving dinners provided for them on these anniversaries. On Thanksgiving, we had religious exercises; on Christmas, a tree well loaded with presents, and on the Fourth of July a suitable celebration.

We have also, through the kindness of the Southern Illinois Fair Association, of Anna, and the Union County Fair Association, of Jonesboro, been able to permit many of the patients to attend the exhibitions of these societies.

Added to all these sources of diversion, there has been a constant daily effort to make the wards as pleasant as possible with music, games, books, papers, etc.

In another place we acknowledge favors received from friends outside, who, by their contributions and assistance, have done much to aid us in this part of the treatment of our patients, and we hope they will not grow "weary in well-doing," but be even more generous in the future.

THE WATER SUPPLY.

As stated in the last biennial report, a contract was let to John Davis & Co., of Chicago, to put up a duplex Worthington pump, with necessary pipes, hose and fittings, to be used in case of fire. This work was nearly completed when, on the twenty-eighth of October, 1878, the pond which was relied upon to furnish water failed entirely, and the only other source at the time available, the cistern below the pond, furnished only about one-sixth of the necessary quantity. By working day and night, Mr. Davis was able to pump water from the small pond, where the fire-pump was erected, on the morning of the fifth of November. This gave temporary relief. In the meantime, a contract was let to the same party, at a called meeting of your board, to put up another duplex Worthington pump at the "big spring," thirteen hundred and fifteen feet southeast of the building, with necessary pipes to connect with that bringing the water from the elevated tank, for the sum of one thousand four hundred and twenty-nine dollars, which amount was paid out of the ordinary fund. This work was completed and water was taken from that spring on the ninth of December following. We have since had a plentiful supply of water, much superior as to healthful qualities to that previously in use.

The appropriation subsequently made by the general assembly for water supply was used to perfect the system and to replace defective pipes in the building, and to carry the supply to the stables, piggery and garden.

The works at the old pond, which now contains about twelve million gallons, are kept in order and ready for use should any emergency arise to make it necessary to resort to that source again.

The old upright boiler, which was used at the "big spring," having become so defective and leaky as to be unfit for use, a contract was let to John O'Brien, of St. Louis, to put up a new one, forty-four inches in diameter, and fourteen feet long, for the sum of seven hundred and twenty-three dollars. This boiler is on the way, and will soon be placed in position.

The present state of the water supply is such as to give a feeling of security as to the future, so far as quantity is concerned.

The "big spring" flows from a cavern in the rock, which runs in a westerly direction from the outlet. Along its course are a large number of openings through the rocky roof, through which the clay soil has washed, forming grottoes or hollows in the surface of the ground, vernacularly called "sink-holes." The surface wash into these "sink-holes" in times of rain and wet weather, makes the water very muddy and unfits it for the general use we are obliged to make of it.

THE SEWER.

The last general assembly made an appropriation of fifteen hundred dollars for extending the sewer. As soon as the appropriation became available work was commenced, and an extension of about fifteen hundred feet leading to the creek east of the building was constructed. This carries the outlet so far away that the gasses arising from the sewage are not perceptible at the building. To this may be attributed, to a great extent, the good health of the inmates during the past year. Those members of the visiting committees from the general assembly who interested themselves in getting this appropriation will be gratified to know that the result has fully answered their expectations.

The new barn and carriage-house, which were under way at date of last report, have been finished and occupied.

The old barn, for the removing and refitting of which an appropriation was made by the last assembly, has been located on a line north of the new one, and fitted up as a bank barn, with stabling for young cattle, a vegetable cellar, and large hay lofts. The barn room is now ample for the present needs of the institution, and the buildings are so located that the drainage can never contaminate the water at the "big spring," or any water used at the hospital.

A chicken-house and park, and a piggery, sixteen by one hundred feet, have been constructed, principally out of surplus and old materials, the work being mostly done by the employés, and without any extra expense to the state. The piggery is large enough to keep the number of swine necessary to consume the offal from the kitchen, and will answer all present purposes.

The appropriation for a new kitchen was expended in the erection of a building thirty by fifty feet, with an ell fifteen by thirty-five. The floor is of marble tile laid in cement. No material of a perishable nature has been used below the frame-work of the roof, except the wood necessary for doors and windows. It is furnished with a new double twelve-foot French range, steam, gas, hot and cold water, complete.

The kitchen could not have been built and fitted as it now is, for the amount of the appropriation, but by the closest economy. The brick were made on the premises, principally by the work of the patients, at an expense not exceeding one dollar per thousand to the state.

We have had great satisfaction in the occupancy of this building. The improvement in the facilities for carrying on the work in the culinary department can hardly be over estimated.

REPAIRS.

The walls, ceilings, and inside wood work of the north wing have been painted and greatly improved in appearance and healthfulness. The old kitchen is being refitted for a dining-room for employés. This has long been needed and will be a great convenience to the hospital.

Some rooms in the basement, used for stores, have been plastered and paved with asphalt blocks. The laundry, where the concrete originally put down had nearly worn out, has been paved with the same material. Other paving is now in progress where the wood floors have so far decayed as to require renewing. These blocks for paving purposes are very durable, and possess an advantage over other kinds of pavement, from the fact that wherever it might become necessary to remove the paving for any purpose, the blocks are in no way damaged by removal and can be replaced with little trouble.

THE GROUNDS.

The appropriation for the improvement of the grounds has been expended, principally, in grading and gravelling the roads, driveways and walks, according to the plans of Messrs. Cleveland & French, landscape architects and engineers, of Chicago. Much of the ground in front of the building has been graded, and the mason work for a fountain-basin, twenty feet in diameter, partly completed. The road to Anna has also been re-covered with a three inch layer of gravel. This will give a hard, durable road-bed, that, with occasional repairs—which can be made without extra expense—will be permanent. I cannot say as much of the two bridges; they are showing evidences of decay, and will need rebuilding in two or three years, —or what would be better, replacing with culverts of masonry. The latter would undoubtedly be the most economical, in the long run. Along either side of the main driveway flower-beds have been prepared, and roses and other perennials and annuals planted.

REQUIREMENTS.

It will require an appropriation of five thousand dollars per annum, for the next two years, for the purpose of improvement and repairs. This is only three-fourths of one per cent. of the cost of the property. It cannot be kept in good repair, and such minor improvements made as are needed from time to time, with a less amount of money.

For the improvement of the grounds, it will require fifteen hundred dollars per annum. With this amount, during the next two years the grounds can be put in a condition to be a credit to the state, and a source of pleasure and enjoyment to the unfortunate inmates that will well repay the outlay.

SETTLING BASIN AND FILTER.

A properly constructed settling basin and filter for the water supplied to the house is very much needed for reasons set forth on a preceding page. It should be constructed so as to be permanent, and of sufficient capacity to furnish at least forty thousand gallons of pure water each twenty-four hours. This will cost, according to the plan and estimate of the architect, herewith submitted, ten thousand dollars. As no one thing can be of greater importance to the welfare of the inmates of a large institution than a plentiful supply of pure water, the amount asked for this purpose cannot, I believe, be considered unreasonable.

VENTILATION.

The bad ventilation of the north wing has been a subject of thought for some years. Last year an experiment was tried to force the foul air out of the building by putting in a small Sturdevant exhaust and blower at the angle of the foul air duct under the centre building, which was run by a small upright engine borrowed from the carpenter shop. This worked successfully so far as the first and second stories were concerned, but it had not capacity to produce a sufficient change of the air above the second floor. A larger fan, run by an engine of suitable power, will no doubt accomplish the purpose, when there is sufficient ingress of pure air to replace that withdrawn. Upon examination it is found that the flues for the ingress of fresh air have an area of only thirty-three inches, while those for the escape of the foul air have an area of eighty-four inches. The fan provided for forcing fresh air into the building cannot be run because of want of power in the engine, which only has a working capacity sufficient to run the machinery in the laundry with the steam pressure we feel safe in carrying.

Provision for increasing influx of fresh air will be necessary before the ventilation can be made satisfactory. It may, and probably will, require additional fresh air flues, even when the fan can be run.

For putting in an exhaust and blower with a suitable engine will require one thousand dollars.

NEW ENGINE.

The engine now in use is too small to perform the work which was intended. We were obliged to disconnect the fans for both the north and south wings about two years ago, to enable us to run the machinery in the laundry. It will require, for supplying the hospital with a suitable engine, according to the estimate of John Davis & Co., the sum of twenty-six hundred and thirty-five dollars.

REFRIGERATING HOUSE.

For the preservation of perishable articles of food, no addition is more needed than a suitable refrigerating house. The necessity for this is so obvious in this climate that it seems superfluous to present arguments in its favor. In the preserving of meats, butter, eggs, etc., which could be procured when the market was low, it would soon pay for itself. The architect estimates the cost, according to the accompanying plan, at six thousand dollars.

EXTENSION AND MACHINERY OF SHOPS.

The carpenter and machine shops are altogether too small. We need an extension of thirty by fifty feet, two stories high (the carpenter shop being over the machine shop). This would give room for the use of some machinery, of which the institution is much in need. For the erection of such extension and procuring necessary machinery, an appropriation of twenty-five hundred dollars will be required.

FENCES.

The fences have been an eye-sore to visitors at the institution, and have evoked criticism and ridicule. The old rail fence enclosing state land on east side of road to Anna, at least, should be replaced with a suitable board or picket fence that will be in keeping with other improvements. For this purpose a sum of five hundred dollars will be required.

The grounds about the building are also enclosed by a common rail fence. We are trying to raise a hedge of Osage-orange to take its place, but some time will elapse before the hedge will be sufficiently strong for that purpose.

LIGHTNING RODS.

The building was struck by lightning on the sixth day of August, 1879, in three places, doing some damage to the ventilators of the south wing and to the windows of the north wing. It was again struck on the twentieth day of August, 1880, damaging to considerable extent one of the ventilators of the south wing.

The building being covered with a mansard roof, I have been in much fear that fire might originate in this way and cause the destruction of the property. It will require one thousand dollars to put rods upon the buildings that would afford protection, and I shall not feel that I have discharged my duty without urging that this appropriation be made.

ORDINARY EXPENSES.

The amount of eighty-eight thousand dollars per annum will be barely sufficient to carry the hospital through another biennial period. This estimate, on a basis of five hundred patients (and the number will not be less), presupposes an annual expense *per capita* of one hundred and seventy-seven dollars, or a fraction over forty-eight cents and two mills per day. It has been in contemplation to change some of the rooms so as to accommodate twenty-five more patients. Should this be determined upon, it will require ninety-two thousand dollars per annum.

SUMMARY OF APPROPRIATIONS ASKED.

Annual:

Ordinary, per annum	$88,000 to $92,000 00
Improvements and repairs, per annum	5,000 00
Improvement of grounds, per annum	1,500 00
Special :	
New engine	2,635 00
Refrigerating house	6,000 00
Extension of shops and machinery	2,500 00
Settling basin and filter	10,000 00
Fencing	500 00
Lightning rods	1,000 00
Fan and engine for ventilation	1,000 00

Attached hereto are tables showing the financial condition of the hospital, an itemized statement of expenses, inventory, and others, which explain themselves, and to which your attention is directed.

ACKNOWLEDGMENTS.

The hospital has been the recipient of the following contributions for the benefit of the patients, for which we return thanks to the donors:

The Farmer and Fruit Grower, Anna.
Union County News, Anna.
The Southern Illinois Advocate, Anna.
Jackson County Era, Murphysboro.
Greenville Advocate, Greenville.
Greenville Sun, Greenville.
The Plain Dealer, Sparta.
The Centralia Democrat, Centralia.
The Radical Republican, Cairo.
The DuQuoin Tribune, DuQuoin.
Shelby County Leader, Shelbyville.
Newman Independent, Newman.
Marion County Herald, Salem.
Shelbyville Union, Shelbyville.
Der Tägliche Stern (daily), Belleville.
The Nashville Democrat, Nashville.
The Legal News, Chicago.
Richland County Republican, Olney.
Cairo Weekly Bulletin, Cairo.
Shawneetown News, Shawneetown.
Kinmundy Independent, Kinmundy.
Freeport Journal, Freeport.
Denver Weekly Times, Denver, Colorado.
Harrisburg Chronicle, Harrisburg.
N. Y. Staats Zeitung, New York.
Carmi Times, Carmi.
Chicago Ledger, Chicago.
Fairfield Democrat, Fairfield.
The Sullivan Progress, Sullivan.
Effingham Democrat, Effingham.

Effingham Republican, Effingham.
Ovid Independent, Ovid, New York.
Effingham Volksblatt, Effingham.
Illinois Volksblatt, Nashville.
The Independent, Murphysboro.
The Oakland Herald, Oakland, California.
The Centralia Sentinel, Centralia.
Argus-Journal, Cairo.
The McLeansboro Times, McLeansboro.
Merchants' Reporter, Chicago.
Saline County Sentinel, Harrisburg.
New York Weekly Tribune, New York.
Wayne County Press, Fairfield.
Danville Commercial, Danville.
Shelbyville Democrat, Shelbyville.
Danville News, Danville.
Valley Clarion, Chester.
Alton Telegraph, Alton.
Mattoon Commercial, Mattoon.
Edwardsville Republican, Edwardsville.
Olney News, Olney.
Jonesboro Gazette, Jonesboro.
The Sentinel, Red Bluffs, California.
Perry County Democrat, Pinckneyville.
The Examiner, San Francisco, California.

To Mrs. Willard, of Jonesboro, and Mrs. Bell, of Cobden, each, for a lot of house plants.

An amateur society, of Cobden, for an evening's entertainment.

An amateur company, of Anna, for the rendering of "Pinafore."

The Stultz Dramatic Company, for a play rendered.

And to several persons from Anna for assistance in the amusements, from time to time.

Also, the following named persons, for contributions for the Christmas trees: McNeil & Higgins, F. McVeagh, Burly & Tyrrell, Janson, McClurg & Co., C. P. Kellogg & Co., Field, Leiter & Co., Jno. Davis & Co., Gale & Bloekie, Markly, Alling & Co.—all of Chicago; Barcley Bros., of Cairo; John E. Dufkin, M. Ussery, W. H. Willard, A. D. Bohannan and Kirkham & Brown, of Anna.

VISITORS.

That the public might have an opportunity to know something of the condition of the hospital and its inmates, the afternoon of every Tuesday, Thursday and Saturday have been designated as regular visiting days. Occasionally, at other times, for some special reasons, visitors have been admitted.

They have been conducted through all the wards except those in which the patients would be likely to be injured by the excitement occasioned by the presence of strangers.

On the first of October, 1878, a visitor's register was opened, which contains the names of three thousand three hundred and forty-eight persons who have visited the institution during the past two years. In the convalescent wards, particularly, these visits are beneficial to the patients, by diverting the thoughts and helping to relieve the monotony of hospital life.

CHANGES.

On first day of November, 1878, Dr. E. D. Converse resigned the position of assistant physician. His place was filled by the selection of Dr. L. E. Stocking, a well educated physician, of Anna, Ill. On the thirty-first day of August, 1879, Dr. F. W. Mercer resigned the place of first assistant, for the purpose of entering upon private practice in the city of Chicago. Dr. Mercer had been connected with the hospital from the first, and had established a reputation as an excellent physician and a worthy man. His place was filled by Dr. W. W. Hester, who had had many years of experience in the treatment of the insane in a neighboring state.

The medical officers above named, and the other officers, and the employés, by their earnest devotion to duty and hearty coöperation in all plans for the welfare of the patients, have advanced the interest of the hospital, lightened my labors, and shown themselves worthy of the trusts assigned them.

To you, gentlemen, I feel under deep obligations for your confidence, advice and support in the discharge of the arduous and responsible duties of my position. And I trust, that in working out the problem of the best treatment and care of these unfortunates, this hospital will not fall behind the foremost in the country.

H. WARDNER, M.D.,
Superintendent.

DIET LIST FOR WEEK.

SUNDAY.

Breakfast.	Dinner.	Supper.
Beefsteak, Potatoes, Bread, Butter, Coffee and tea.	Baked pork and beans, Bread, Potatoes, Pickles, Pie.	Bread, Butter, Tea, Syrup.

MONDAY.

Boiled meat, Potatoes, Bread, Butter, Coffee and tea.	Roast meat, Potatoes, Beets or turnips, Bread.	Brown bread, Butter, Peach or apple sauce, Tea.

TUESDAY.

Hash, Bread, Butter, Coffee and tea.	Roast meat, Potatoes, Bread, Bread pudding.	Bread, Butter, Cheese, Tea.

WEDNESDAY.

Bacon and ham, Potatoes, Bread, Butter, Coffee and tea.	Bean soup, with Beef or mutton, Potatoes, Bread, Onions.	Bread, Butter, Apple sauce, Tea.

THURSDAY.

Boiled meat, Potatoes, Bread, Butter, Coffee and tea.	Fried steak, Potatoes, Bread, Rice, with sauce.	Bread, Butter, Ginger bread, Tea.

FRIDAY.

Salt fish, Bread, Butter, Potatoes, Coffee and tea.	Boiled ham, Hominy, (For one wing, alternating.) Fresh fish, Potatoes, (For other wing, alternating.) Bread, Syrup.	Mush, Bread, Syrup, Butter, Tea.

SATURDAY.

Boiled meat, Potatoes, Bread, Butter, Coffee and tea.	Bean soup, with Boiled meat, Potatoes, Bread, Syrup.	Corn bread, Bread, Butter, Tea.

In addition to the above, fruits and fresh vegetables are freely supplied to the whole house during the season for them. There is also a special diet list for the sick and feeble, which is modified daily, according to needs of the patients.

Farm Produce.

3,165	bushels corn, average value for two years at 37½c	$1,188 25	
741	bushels oats, average value for two years at 33c	244 35	
643	bushels potatoes, average value for two years at 62c	399 15	
137	tons hay, average value for two years at 9.87 10/17c	1,333 00	
1,228	bushels apples, average value for two years at 43⅔c	536 30	
45	bushels apples, average value for two years at 50c	22 50	
70½	loads straw, average value for two years at $1.00	70 50	
1,400	shocks corn fodder, average value for two years at 15c	210 00	
15,883	gallons milk	2,819 32	
437	bushels sweet potatoes, at 44½c	194 80	
			$7,018 17
	RECEIPTS FROM SALES OF LIVE STOCK.		
	Pigs	$12 00	
	Calf	6 00	
			18 00
	MISCELLANEOUS RECEIPTS.		
	Hides	$65 65	
	Old house	10 00	
			75 65
	VALUE OF STOCK SLAUGHTERED FOR USE OF HOSPITAL.		
6,799	lbs. pork, at 3c	$203 97	
430	lbs. veal, at 7c	30 10	
1,113	lbs. veal, at 6c	66 18	
3,258	lbs. beef, at 6.15	217 23	
5,827	lbs. pork, at 4½c	262 21	
			779 69
			$7,891 51

Garden Produce.

2,787	bushels onions, at 5c	$139 35
486	bushels onions, at 50c	243 00
20	bushels onion sets, at 50c	10 00
129	crates lettuce	56 40
70	crates spinach at 38½c	27 00
17	bbls. spinach, at $1.50	25 50
400	crates pie plant, at 30c	120 00
44	bunches pie plant, at 10c.	4 40
28	bbls. pie plant, at $1.00	28 00
1,701	bunches radishes, at 5c	85 05
215	bunches asparagus, at 10c	21 50
8,153	quarts strawberries, at 9c	733 77
122	bushels peas, at 94c	114 67
94	bushels string beans, at 77 82/94c	73 20
115	bushels beets, at 42⅝c	49 00
10	bbls. beets, at 75c	7 50
178	bushels cucumbers, at 65½c	116 60
4,190	cabbage heads, at 6 6/10c	279 22
948	squashes, at 3⅔c	34 44
30	bushels early turnips, at 75c	22 50
324	bushels late turnips, at 39¾c	127 20
661	bushels tomatoes, at 49c	325 20
16	bushels peppers, at 44c	7 00
330	bushels parsnips, at 45⅓c	149 70
158	bushels carrots, at 42⅔c	67 40
188	bushels ruttabagas, at 43⅞c	82 65
785	bunches celery, at 5¾c	44 95
448	bushels sweet corn, at 56c	251 30
17	bushels butter beans, at $1.00	17 00
80	pounds sage, at 5c	4 00
50	pounds sage, at 10c	5 00
6	loads pumpkins, at $3.00	18 00
2	bushels okra, at $1.00	2 00
155	bushels sweet potatoes, at 40c	62 00
		$3,354 50

·

MATRON'S REPORT.

Articles made in Sewing Room, for two years ending September 30, 1880.

Aprons	304	Sacks, clothes	24
Bonnets	35	Suits, duck	11
Bibbs	15	Sheets	514
Chemises	285	Skirts, flannel	302
Caps, cook's	57	Skirts, white cotton	75
Curtains	32	Skirts, colored	21
Dresses	954	Table cloths	121
Dresses, night	86	Towels, roller	277
Drawers, men's	118	Ticks, bed	34
Drawers, women	28	Ticks, pillow	40
Dish towels	710	Ticks, mattress	51
Napkins	139	Waists, flannel	88
Shirts, plain cotton	454	Waists, cotton	134
Shirts, flannel	29	Waists, duck	3
Shirts, under	22		

ARTICLES MENDED IN SEWING ROOM,

Coats	701	Drawers	1,263
Vests	588	Socks	1,532
Pants, pairs	1,255	Shirts	1,402
Overjackets	23	Waists	6
Bibbs	34	Caps	21

FRUITS CANNED AND PRESERVED.

Apples, pickled	Galls.	6	Apple jelly	Glasses 955
Tomatoes	"	500	Cranberry jelly	" 40
Chow-chow	"	300	Apples, preserved	Galls. 48
Cucumbers, pickled	"	480	Vinegar made	" 796

FINANCIAL STATEMENT

Of the Receipts and Expenditures of the Illinois Southern Hospital for the Insane, for the fiscal year from October 1, 1878, to September 30, 1879.

ORDINARY EXPENSE.

Dr.

1878.		
October 1..... To balance..	$32,977 38	
October 1..... " appropriation for quarter ending December 31....	21,250 00	
1879.		
January 1..... To appropriation for quarter ending March 31........	21,250 00	
April 1........ " appropriation for quarter ending June 30..........	21,250 00	
July 1......... " appropriation for quarter ending September 30....	16,500 00	
September 30. " counties, for clothing...............................	4,634 94	
September 30. " individuals, for clothing............................	3,343 44	
September 30. " sales of live stock.................................	32 00	
September 30. " sales of farm produce..............................	265 96	
September 30. " all other sources	1,496 18	

Cr.

1879.		
September 30. By indebtedness incurred on account of—		
Attendance—salaries and wages		$30,463 10
Food..		27,577 09
Clothing, bedding, etc		3,305 40
Laundry supplies		773 70
Fuel		3,019 51
Light		1,471 38
Medicines and medical supplies.......................		2,150 32
Freight and transportation		5,408 00
Postage and telegraphing.............................		446 17
Books and stationery.................................		395 03
Printing and advertising.............................		241 60
Music and amusements		546 91
Instruments and apparatus............................		33 28
Household expenses...................................		862 12
Furniture		2,055 81
Buildings, improvements and repairs		195 71
Tools		19 47
Machinery, etc		1,629 02
Farm, garden, stock, grounds, roads and fences.......		1,551 29
Legal expenses.......................................		356 60
Shop expenses..		75 84
Burial expenses......................................		131 00
Expenses not classified		13 20
By balance		39,678 35
	$122,399 90	$122,399 90
1879.		
October 1..... To balance, in hands of R. B. Stinson, treasurer......	$27,961 51	
October 1..... " " in state treasury, undrawn	16,500 00	
October 1..... " " in hands of W. N. Mitchell, ex-treasurer.	498 88	
	$44,960 39	
Less indebtedness outstanding, which includes $1,020.68 of water supply $5,240 01		
Less orders unpaid 42 03		
	5,282 04	
September 30. To balance...	$39,678 35	

Financial Statement—*Continued.*

IMPROVEMENTS AND REPAIRS.

Dr.

1878. October 1	To balance	$4,485 99	
1879. July 1	To amount of appropriation	4,000 00	

Cr.

1879. September 30.	By indebtedness incurred on account of—		
	Attendance		$1,860 17
	Freight		21 15
	Household expense		51 85
	Furniture		31 13
	Improvements and repairs		3,599 27
	Farm, garden, etc.		96
	Machinery		1,423 33
	Tools		80 57
	By balance, in state treasury, undrawn		1,417 56
		$8,485 99	$8,485 99
1879. October 1	To balance	$1,417 56	

IMPROVEMENT OF GROUNDS.

Dr.

1878. October 1	To balance	$212 87	
1879. July 1	To amount of appropriation	1,000 00	

Cr.

1879. July 1.	By indebtedness incurred on account of—		
	Attendance		$295 12
	Improvement and repairs		160 50
	Farm, garden, etc		5 00
	By balance, in state treasury, undrawn		752 25
		$1,212 87	$1,212 87
1879. October 1	To balance	$752 25	

CARPENTER SHOP.

Dr.

1878. October 1	To balance	$209 39	

Cr.

1879. September 30.	By indebtedness incurred on account of—		
	Machinery		$209 39
		$209 39	$209 39

FIRE, PUMP AND HOSE.

Dr.

1878. October 1	To balance	$1,800 00	

Cr.

1879. September 30.	By indebtedness incurred on account of—		
	Fixtures, hose and carriage		$1,800 00
		$1,800 00	$1,800 00

Financial Statement—*Continued*.

ROTARY OVEN.

Dr.

1878.		
October 1..... To balance..	$306 40	

Cr.

1879.		
September 30. By indebtedness incurred on account of—		
Attendance......................................		$20 00
Improvement and repairs		79 94
By amount in state treasury, undrawn (lapsed)........		206 46
	$306 40	$306 40

NEW KITCHEN.

Dr.

1879.		
July 1......... To amount of appropriation.........................	$3,000 00	

Cr.

1879.		
September 30. By indebtedness incurred on account of—		
Attendance......................................		$100 50
Freight..		1 75
Improvement and repairs		1,150 33
By balance, in state treasury, undrawn...............		1,747 42
	$3,000 00	$3,000 00
1879.		
October 1..... To balance	$1,747 42	

WATER SUPPLY.

Dr.

1879.		
July 1......... To amount of appropriation	$2,500 00	

Cr.

1879.		
September 30. By indebtedness incurred on account of—		
Attendance......................................		$27 00
Freight..		2 30
Improvement and repairs........................		1,786 38
By balance in state treasury, undrawn		684 32
	$2,500 00	$2,500 00
1879.		
October 1..... To balance	$684 32	

REMOVAL OF BARN.

Dr.

1879.		
July 1......... To amount of appropriation	$1,000 00	

Cr.

1879.		
September 30. By indebtedness incurred on account of—		
Improvement and repairs........................		$511 00
By balance, in state treasury, undrawn		489 00
	$1,000 00	$1,000 00
1879.		
October 1..... To balance	$489 00	

Financial Statement—*Continued.*

	EXTENDING SEWER.		
	Dr.		
1879. July 1.........	To amount of appropriation...............................	$1,500 00	
	Cr.		
1879. September 30.	By indebtedness incurred on account of—		
	Attendance..	....	$150 67
	Improvement and repairs...............................	....	983 17
	By balance in state treasury, undrawn...............	....	366 16
		$1,500 00	$1,500 00
1879. October 1.....	To balance..	$366 16	

FINANCIAL STATEMENT

Of the Receipts and Expenditures of the Illinois Southern Hospital for the Insane, for the fiscal year from October 1, 1879, to September 30, 1880.

	ORDINARY EXPENSE.		
	Dr.		
1879.			
October 1.....	To balance...	$39,678 35	
October 1.....	Appropriation for quarter ending December 31....	16,500 00	
1880.			
January 1....	To appropriation for quarter ending March 31..........	16,500 00	
April 1.......	Appropriation for quarter ending June 30..........	16,500 00	
July 1.........	Appropriation for quarter ending September 30 ...	22,500 00	
September 30.	Counties, for clothing.....................................	4,686 36	
September 30.	Individuals, for clothing	2,791 38	
September 30.	Sales of live stock.......................................	18 00	
September 30.	All other sources...	898 91	
	Cr.		
1880.			
September 30.	By indebtedness incurred on account of—		
	Attendance (salaries and wages)		$30,163 07
	Food		30,882 63
	Clothing, bedding, etc....................................		9,313 70
	Laundry supplies		947 59
	Fuel		2,914 95
	Light		2,316 69
	Medicines and medical supplies.........................		1,751 73
	Freight and transportation ·········		5,805 27
	Postage and telegraphing................................		312 88
	Books and stationery.....................................		314 68
	Printing and advertising		180 80
	Music and amusements		590 09
	Instruments and apparatus..............................		34 33
	Household expenses.......................................		1,011 35
	Furniture		2,297 82
	Buildings, improvements and repairs		2,180 79
	Tools		106 75
	Machinery, etc...		161 35
	Farm, garden, stock, grounds, roads and fences...		1,420 57
	Legal expenses...		36 80
	Shop expenses		90 30
	Burial expenses		94 00
	Expenses not classified		62 56
	By balance..		27,082 31
		$120,073 01	$120,073 01
1880.			
October 1.....	To balance, in hands of R. B. Stinson, treasurer......	$13,567 30	
	Bills outstanding...	9,779 20	
		$3,788 10	
	Balance in state treasury, undrawn	23,294 21	
	To balance...	$27,082 31	

Financial Statement—*Continued.*

IMPROVEMENTS AND REPAIRS.

Dr.

1879. October 1.....	To balance...	$1,417 56	
1880. July 1.........	Amount of appropriation	4,000 00	

Cr.

1880. September 30.	By indebtedness incurred on account of—		
	Attendance	....	$913 84
	Freight and transportation.....................	....	18 46
	Household expenses.............................	....	19 85
	Furniture......................................	....	1 20
	Building, improvements and repairs............	....	2,343 88
	Farm, garden. stock, etc.......................	....	6 90
	Machinery and fixtures.........................	....	82 37
	Tools	....	11 75
	By balance, in state treasury, undrawn...........	....	2,019 31
		$5,417 56	$5,417 56
October 1.....	To balance	$2,019 31	

IMPROVEMENT OF GROUNDS.

Dr.

1879. October 1.....	To balance...	$752 25	
1880. July 1.........	Amount of appropriation........................	1,000 00	

Cr.

September 30.	By indebtedness incurred on account of—		
	Attendance.....................................	....	$1,167 38
	Building, repairs, etc..........................	....	72 55
	Farm, garden, stock, etc.	....	162 44
	By balance, in state treasury, undrawn...........	....	349 88
		$1,752 25	$1,752 25
October 1.....	To balance	$349 88	

NEW KITCHEN.

Dr.

1879. October 1.....	To balance...	$1,747 42	

Cr.

1880. September 30.	By indebtedness incurred on account of—		
	Attendance	....	$129 49
	Building, repairs, etc	....	1,117 98
	Machinery and fixtures.........................	....	488 00
	By balance, in state treasury, undrawn	....	11 95
		$1,747 42	$1,747 42
October 1.....	To balance...	$11 95	

WATER SUPPLY.

Dr.

1879. October 1.....	To balance...	$684 32	

Cr.

1880. September 30.	By indebtedness incurred on account of—		
	Attendance	....	$47 24
	Freight and transportation	....	2 25
	Furniture	....	16 00
	Building, repairs, etc	....	344 75
	By balance, in state treasury, undrawn	....	274 28
		$684 32	$684 32
October 1.....	To balance...	$274 28	

Financial Statement—*Continued*.

	REMOVAL OF BARN.		
	Dr.		
1879. October 1.....	To balance..	$489 00	
	Cr.		

1880.
September 30.
	By indebtedness incurred on account of—
	    Attendance ....
	    Freight and transportation ....
	    Building, repairs, etc ....
		....
		....
		....
			$35 00
			1 25
			452 75

		$489 00	$489 00
	EXTENDING SEWER.		
	Dr.		
1879. October 1.....	To balance	$366 16	
	Cr.		

1880.
September 30.
	By indebtedness incurred on account of—
	    Attendance ....
	    Building, repairs, etc..............................
	By balance, in state treasury, undrawn ....
		....
		....
		....
			$44 53
			84 90
			236 73

		$366 16	$366 16
October 1.....	To balance.	$236 73	

ITEMIZED STATEMENT

Of the kind, quantity and cost of all articles purchased for the Institution, during the fiscal years 1878 and 1880, from October 1, 1878, to September 30, 1880.

ALL FUNDS CONSOLIDATED.

Item.	Measure	1878-9. Am't.	Cost.	1879-80. Am't.	Cost.	1878 and 1880. Am't.	Cost.
ATTENDANCE.							
Salaries			$8,293 52		$8,349 21		$16,642 73
Wages of employes			21,697 78		21,417 05		43,114 83
LABOR, NOT ON PAY ROLL.							
Bricklayers	Days	122¼	160 37		191 05		351 42
Carpenters and joiners	"	194	640 19		463 80		1,103 99
Cleaning house	"		3 90				3 90
Day laborers	"	470¼	625 92		468 25		1,094 17
Grading	"		6 75				6 75
Labor with team	"				890 58		890 58
Painting	"	310	598 11		332 38		930 49
Plasterers	"	294¾	583 87		110 00		693 87
Roofing	"	1	3 00				3 00
Steam fitting	"	9	27 00				27 00
Threshing	"		21 15		13 23		34 38
Minister	"	51	255 00		265 00		520 00
Total attendance			$32,916 56		$32,500 55		$65,417 11
FOOD.							
BREADSTUFFS.							
Baking powder	Pounds	322	$98 35	1,120	$310 30	1,442	$408 65
Baking soda	"	43	3 58	57	4 70	100	8 28
Carbonate of ammonia	"	5½	1 90	7	3 20	12½	5 10
Corn starch	"	1,350	98 10	1,280	91 10	2,630	189 20
Cracked wheat	"	25	3 65			25	3 65
Crackers	"	993	52 56	2,506	133 29	3,499	185 85
Cream tartar	"			1	75	1	75
Flour, graham	"	2,900	64 25	4,100	112 75	7,000	177 00
Flour, wheat	"	144,226	3,541 05	150,050	4,276 70	294,276	7,817 75
Hops	"	46	8 20	50	25 00	96	33 20
Malt	"	20	82	20	1 00	40	1 82
Meal, corn	"	6,200	56 40	9,000	91 40	15,200	147 80
Meal, oat	Barrels	6	45 43	16	81 27	22	126 70
Sago	Pounds	122½	7 35			122½	7 35
Tapioca	"	122	7 93			122	7 93
Yeast	"				15 25		15 25
Yeast powder	"	25	6 25			25	6 25

Itemized Statement—*Continued.*

Item.	Measure	1878-9. Am't.	Cost.	1879-80. Am't.	Cost.	1878 and 1880. Am't.	Cost.
MEATS, ETC.							
Fresh.							
Beef, quarters	Pounds..	145,568	$8,947 61	141,890	$9,144 74	287,458	$18,092 35
Lamb	"	33	3 30			33	3 30
Mutton	"			116	6 38	116	6 38
Sausage	"	384	26 92	45	3 60	429	30 52
Beef, canned	"	24	9 40	96	28 08	120	37 48
Salt.							
Beef	Pounds.	3,600	177 00			3,600	177 00
Pork	"	6,200	291 85	41	492 00	14,400	783 85
Smoked.							
Beef, dried	Pounds.			100	7 50	100	7 50
Breakfast bacon	"	1,770	136 71	331	29 79	2,101	166 50
Hams	"	4,062	335 96	1,757	175 27	5,819	511 23
Shoulders	"	344	17 20			344	17 20
Sides	"	725	55 26			725	55 26
Fish.							
Fresh	Pounds.	4,030	411 94	4,779	508 60	8,809	920 54
Salt	"	10	47 00			10	47 00
Canned	Cans....	24	9 25			24	9 25
Oysters	"	155	43 02	26	9 95	181	52 97
Oysters, cove	"	12	1 50	504	65 70	516	67 20
Codfish	Pounds..	95	5 66			95	5 66
Herring	Boxes...	12	2 40			12	2 40
Mackerel	Barrels..	15	150 00	96	307 50	111	457 50
Whitefish	Pounds..	1,200	43 20			1,200	43 20
Poultry.							
Chickens, live	Number	3,370	585 43	2,822	487 54	6,192	1,072 97
Ducks	"	33	3 30	30	4 66	63	7 96
Geese	"	4	2 00	24	9 50	28	11 50
Turkeys. live	"	7	7 25	52	26 10	59	33 35
Turkeys, dressed	Pounds.	736	62 94	574	45 92	1,310	108 86
Geese, dressed	"	71	4 97			71	4 97
Miscellaneous.							
Gelatine	Pounds.		26 20				26 20
Lard	"	2,998	321 94	4,756	398 67	8,754	720 61
Mincemeat	"	628	59 75	331	26 48	959	86 23
VEGETABLES.							
Green.							
Cabbage	Heads...	15	90			15	90
Celery	Bunches	63	3 75			63	3 75
Potatoes, Irish	Bushels.	1,577½	962 78	1,700½	647 34	3,277¾	1,610 12
Potatoes, sweet	"	10	11 50			10	11 50
Potatoes, saratoga	Pounds..	30	6 30			30	6 30
Turnips	Bushels.	4	2 00			4	2 00
Canned.							
Corn	Cans....	96	10 00	144	24 00	240	34 00
Peas	"	10	1 73			10	1 73
Pumpkin	"	48	12 00			48	12 00
Succotash	"	96	16 00			96	16 00
Tomatoes	"	126	14 15	144	19 20	270	33 35
Dried.							
Beans	Pounds.	5,647	135 96	3,122	88 74	8,769	224 70
Hominy	Barrels..	2	5 50	6	23 25	8	28 75
Rice	Pounds.	1,609	109 77	2,078	146 34	3,687	256 11

Itemized Statement—*Continued.*

Item.	Measure	1878-9.		1879-80.		1878 and 1880.	
		Am't.	Cost.	Am't.	Cost.	Am't.	Cost.
Pickles.							
Green	Barrels	10	$56 00			10	$56 00
Capers	Bottles	12	4 50			12	4 50
Chow-chow	"	12	5 65	48	$16 00	60	21 65
Cucumbers	"	24	9 00			24	9 00
Peppers	"	24	2 20			24	2 20
Sauces	"	36	14 75	36	11 65	72	26 40
Vinegar	Gallons	298	78 93	640	107 72	938	186 65
FRUIT.							
Green.							
Apples	Bushels	145	116 89	48	28 00	193	144 89
Blackberries	Quarts	108	10 01			108	10 01
Cantelopes	Number	2,501	50 20	1,351	41 09	3,852	91 29
Cranberries	Bushels	6	13 50	3	8 00	9	21 50
Grapes	Pounds	1,520	65 72	1,371	50 84	2,891	116 56
Lemons	Number	1,642	57 85	1,068	38 00	2,710	95 85
Lomons	Boxes			2	11 00	2	11 00
Melons, water	"	258	23 24	879	68 58	1,137	91 82
Oranges	"	348	8 65	6	25	354	8 90
Raspberries	Quarts	93½	9 42	112	5 90	205½	15 32
Canned.							
Apricots	Cans			96	26 00	96	26 00
Blackberries	"	48	3 80			48	3 80
Cherries	"	144	27 60	696	107 50	840	135 10
Peaches	"	352	73 00	384	76 95	736	149 95
Cranberries	"			24	4 20	24	4 20
Plums	"			72	22 35	72	22 35
Quinces	"	6	5 25			6	5 25
Raspberries	"	24	4 50			24	4 50
Huckelberries	"	12	4 50			12	4 50
Dried.							
Almonds	Pounds			1	20	1	20
Apples	"	1,997	84 88	1,055	72 51	3,052	157 39
Blackberries	"			50	5 50	50	5 50
Cherries	"	20	6 00	10	2 00	30	8 00
Cocoanuts	"	24	6 00	96	25 42	120	31 42
Currants	"	50	2 25	275	16 60	325	18 85
Nuts	"	60	6 98			60	6 98
Peaches	"	1,008	33 27	864	102 08	1,872	135 35
Peel, citron	"			12¼	2 45	12¼	2 45
Prunes	"	208	13 57	1,328	106 24	1,536	119 81
Raisins	"		14 80				14 80
Raisins	Boxes	5	10 00	5	12 25	10	22 25
JELLIES, PRESERVES, ETC.							
Jelly	Jars	72	30 25	96	24 08	168	54 33
OTHER PROVISIONS.							
Coffee, etc.							
Cahvey, Ottoman	Pounds	1,274	331 24			1,274	331 24
Chocolate	"			12	4 80	12	4 80
Cocoa	"	100	38 48	168	64 80	268	103 28
Coffee, Rio	"	5,361	744 23	1,067	182 56	6,428	926 79
Coffee, roasted	"	849	117 85	4,346	565 23	5,195	683 08
Tea, green	"	165	65 20			165	65 20
Tea, Japan	"	626	149 32	1,931	570 68	2,557	720 00
Tea, Oolong	"	749	259 30	1,847	583 19	2,596	842 49
Milk, etc.							
Butter	Pounds	17,652½	3,038 96	17,898½	3,928 24	35,551	6,967 20
Cheese	"	1,685	135 66	991	121 78	2,676	257 44
Milk	Gallons	2,908	542 25	4,088	722 65	6,996	1,264 90
Cream	"	7½	5 95	4½	3 52	12	9 47
Eggs	Dozen	12,150½	1,257 85	15,710	1,427 87	27,860½	2,685 72

Itemized Statement—*Continued*.

Item.	Measure	1878-9.		1879-80.		1878 and 1880.	
		Am't.	Cost.	Am't.	Cost.	Am't.	Cost.
Sugar, etc.							
Honey	Pounds..	10	$2 00	5	$1 00	15	$3 00
Molasses	Gallons.	186	65 10			186	65 10
Syrup	"	1,864½	632 51	1,026½	432 42	2,891	1,064 93
Sugar, granulated	Pounds..	3,457	298 17	5,959	608 77	9,416	906 94
Sugar, powdered	"			318	36 57	318	36 57
Sugar, A	"	11,365	931 90	13,988	1,294 09	25,353	2,225 99
Sugar, C	"	4,080	332 00	5,285	438 21	9,365	770 21
Sugar, maple	"	50	5 25	50	7 00	100	12 25
Sugar, brown	"	6,380	461 96	6,157	503 67	12,537	965 63
Spices, etc.							
Allspice	Pounds..	12	2 40	10	2 00	22	4 40
Cinnamon	"	30¼	11 35	45	16 50	75¼	27 85
Cloves	"	9	4 40	10	4 00	19	8 40
Ginger	"	½	25	11	2 35	11½	2 60
Mustard, French	Bottles..	52	4 00	144	9 50	196	13 50
Mustard, ground	Pounds..		39 33	100	20 00		59 33
Nutmegs	"	½	35	5	4 75	5½	5 10
Pepper	"			268	56 56	268	56 56
Miscellaneous.							
Extracts	Bottles..	162	80 75	300	120 00	462	200 75
Ice	Cwt			2,672⁹/10	214 90	2,672⁹/10	214 90
Sage	Pounds..	8½	3 30			8½	3 30
Salt	Barrels .	46	85 50	34	68 10	80	153 60
Total food			$27,577 09		$30,882 63		$58,459 72
CLOTHING, BEDDING, Etc.							
Class 1—*Wearing Apparel.*							
Aprons	Number	12	$3 25			12	$3 25
Belts	"			48	$8 74	48	8 74
Boots	Pairs....	24	51 02	73	180 27	97	231 29
Buskins	"	12	6 84			12	6 84
Cloaks	Number			5	21 70	5	21 70
Coats, linen	"	42	43 23	1	1 75	43	44 98
Coats, woolen	"	93	166 52	292	969 52	385	1,136 04
Collars, linen	"	48	2 85	432	37 43	480	40 28
Collars, paper	Boxes...	60	3 60	200	38 00	260	41 60
Collar buttons	Number			156	3 09	156	3 09
Combs, round	"	48	2 28			48	2 28
Combs, back	"	15	2 99	111	10 04	126	13 03
Corsets	"	36	26 00	24	16 00	60	42 00
Cuffs	Pairs...	24	3 80	72	10 26	96	14 06
Drawers, cotton	"	60	19 00			60	19 50
Drawers, woolen	"	168	66 95	480	186 33	648	253 28
Dusters	Number			1	4 05	1	4 05
Vests	"	72	25 65			72	25 65
Hair-pins	P'kages.	111	3 97			111	3 97
Handkerchiefs	Number	240	14 49	324	27 97	564	42 46
Hats, men's	"	72	47 83	144	68 40	236	116 23
Hats, women's	"	36	10 33	24	7 50	60	17 83
Hats, straw	"	43	6 65			43	6 65
Hoods	"	1	50			1	50
Hose, men's	Pairs	504	43 60	2,476	334 21	2,980	377 81
Hose, women's	"	408	51 53	1,240	179 57	1,648	231 10
Mittens	"	24	21 00	18	8 55	42	29 55
Neckties	Number	156	12 46	204	21 37	360	33 83
Overalls	"	24	22 33			24	22 33
Overcoats	"			12	62 70	12	62 70
Rubber shoes	Pairs...	1	1 30			1	1 30
Scarfs, woolen	Number	12	2 14			12	2 14
Shawls	"			12	11 40	12	11 40
Shirts, cotton	"	149	68 04	144	79 10	293	142 14
Shirts, linen	"			380	342 25	380	342 25
Shoes, men's	Pairs...	174	236 16	112	176 13	286	412 29
Shoes, women's	"	33	22 79	108	161 02	141	183 81
Skirts	Number	46	38 00			46	38 00

Itemized Statement—*Continued.*

Item.	Measure	1878-9. Am't.	1878-9. Cost.	1879-80. Am't.	1879-80. Cost.	1878 and 1880. Am't.	1878 and 1880. Cost.
Slippers, men's	Pairs	227	$193 14	378	$331 80	605	$524 94
Slippers, women's	"	24	14 57	293	243 12	317	257 69
Socks, cotton	"	144	34 20	......	......	144	34 20
Spectacles	"	60	15 25	36	3 84	96	19 09
Suits, men's	Number	95	444 56	126	727 11	221	1,171 67
Suits, women's	"	60	201 50	62	182 15	122	383 65
Suits, linen	"	12	8 55	60	221 25	72	229 80
Suspenders	Dozens	8	20 42	84	41 33	92	61 75
Trousers (pantaloons)	Number	156	182 39	297	617 10	453	799 49
Undershirts	"	137	55 60	510	198 90	647	254 50
Vests	"	83	67 81	244	176 51	327	244 32
Wrappers	"	119	64 24	54	44 88	173	109 12
Class 2—*Beds, Tables, etc.*							
Blankets	Dozens	1¼	34 07	8¼	237 50	9½	271 57
Blankets, rubber	"	......	......	50	62 50	50	62 50
Comforts	Number	......	......	48	35 64	48	35 64
Crash	Yards	69½	35 28	1,245	143 38	1,314½	178 66
Damask	"	......	......	393½	210 63	393½	210 63
Mosquito bars	Number	9	17 55	1	65	10	18 20
Napkins	"	36	9 40	108	10 07	144	19 47
Piano covers	"	1	3 33	......	......	1	3 33
Quilts (spreads)	"	32	49 87	......	......	32	49 87
Sheeting	Yards	......	......	3,706½	540 77	3,706½	540 77
Table-cloths	Number	3	6 00	1	1 90	4	7 90
Table-covers, colored	"	......	......	12	13 40	12	13 40
Ticking	Yards	251¾	32 02	789½	132 04	1,041¼	164 06
Ticks	Number	......	......	50	42 50	50	42 50
Tidies	"	......	......	25	16 58	25	16 58
Toweling	Yards	100	12 86	432	54 00	532	66 86
Towels	Number	12	2 85	230	33 18	242	36 03
Class 3—*Materials.*							
Alpaca	Yards	51¾	18 11	......	......	51¾	18 11
Calico	"	2,380¾	145 18	6,765¼	533 29	9,146	678 47
Cambric	"	......	......	92	6 37	92	6 37
Canvas	"	6	90	84	21 31	90	22 21
Canton flannel	"	......	......	503	49 80	503	49 80
Cheviot	"	929½	81 93	......	......	929½	81 93
Cloth, woolen	"	......	......	6	14 25	6	14 25
Dress goods, cotton	"	250¼	15 35	......	......	250¼	15 35
Dress goods, woolen	"	57¾	9 06	......	......	57¾	9 06
Dress goods, silk	"	......	......	3	3 21	3	3 21
Drilling	"	63½	7 86	97½	8 46	161	16 32
Duck	"	128	18 40	374	58 76	502	77 16
Flannel	"	632¾	154 19	1,042½	270 83	1,675¼	425 02
Gingham	"	115	11 52	806¾	81 67	921¾	93 19
Holland	"	......	......	4	60	4	60
Jeans	"	17	7 92	......	......	17	7 92
Lace	"	3½	3 94	70½	25 65	74	29 59
Linen	"			46½	9 50	46½	9 50
Muslin	"	......	......	3,075	284 88	3,075	284 88
Ribbons	"	10	2 35	......	15 63	......	17 98
Ruching	"	......	......	2	70	......	70
Shirting	"	904	137 90	2,063¾	169 16	......	307 06
Silesia	"	62	10 55	......	......	62	10 55
Tarleton	"	56¼	9 28	131¾	18 55	188	27 83
Velveteen	"	4½	5 18	1½	1 08	6	6 26
Rubber cloth	"	47	28 20	......	......	47	28 20
Class 4—*Findings.*							
Bow-wire		......	......	......	1 87	......	1 87
Binding	Rolls	......	3 03	2	2 05	......	5 08
Binding, stay	"	......	......	......	7 04	......	7 04
Braid	Pieces	11	1 65	1	85	12	2 50
Braid, woolen	"	24	1 30	......	......	24	1 30
Buttons, agate	Gross	......	4 78	24½	14 73	......	19 51
Buttons, brass	Dozens	4	40	12	1 20	16	1 60
Buttons, coat	Gross	2	1 25	......	4 75	2⅔	6 00
Buttons, dress	"	1	1 43	39½	24 67	51½	26 10
Buttons, pearl	"	2	2 38	36	41 08	60	43 46
Button moulds	"	......	......	4	2 85	4	2 85

Itemized Statement—*Continued.*

Item.	Measure	1878-9.		1879-80.		1878 and 1880.	
		Am't.	Cost.	Am't.	Cost.	Am't.	Cost.
Cord	Bolts				$4 00		$1 00
Cotton plush	Yards			57½	15 73	57½	15 73
Lace bunting	"			26	5 70	26	5 70
Oil-cloth	"			1¼	1 25	1¼	1 25
Cotton, darning	Balls	12	$5 66			12	5 66
Cotton, knitting	"	2	83		16 61		17 44
Tassels	No. bxs.			5	4 75	5	4 75
Edgings	Yards			4	1 10	4	1 10
Buckram	"			15	2 23		2 23
Eyelets	Gross	1	2 75			1	2 75
Fringe	Yards			58	12 17	58	12 17
Gimp	"				85		85
Laces, shoe	"	3	7 38	4	2 51	7	9 89
Needles	M	7½	12 49	7	11 65	14½	24 14
Needles, crochet	Dozens			5	3 17	5	3 17
Needles, knitting	"			3	90	3	90
Needles, machine	"	4	2 00		4 90		6 90
Patterns	Number.	20	1 00	77	5 11	97	6 11
Perforated card-board	Sheets	13	1 08	336	14 46	349	15 54
Pins	P'kages.	27	6 97		7 47		14 44
Silk floss	Spools				1 99		1 99
Tape	Dozens.	24	14 25	17	10 02	41	24 27
Thimbles	Number			156	3 49	156	3 49
Thread, cotton	Spools	360	19 50	2,568	124 25	2,928	143 75
Thread, linen	Pounds.	2	2 24		3 67		5 91
Tow	"	5	2 25			5	2 25
Twist, machine	Spools	2	2 00		4 54		6 54
Whalebones	Number.		15				15
Yarn	Pounds.	5½	6 45	49	47 5	54½	54 30
Zephyr	Ounces	12	1 80	78⅞	43 46	90⅞	45 26
Total clothing			$3,305 40		$9,313 70		$12,019 10
LAUNDRY SUPPLIES.							
Bluing	Gallons.			96½	$31 65	96½	$31 65
Flat-irons	Pounds.	15	$14 37			15	14 37
Fluters	Number	1	3 50			1	3 50
Indelible ink	Dozens.	15⅝	31 92	14½	29 34	30⅛	61 26
Lye	Boxes	43	133 50	82	27 30	125	360 80
Potash	"	24	68 48			24	68 48
Soap, hard	Pounds.	9,695	457 69	10,980	541 90	20,675	999 59
Soda, washing	"	297	11 90	746	12 12	1,043	24 02
Starch	"	966	38 24	1,954	101 58	2,920	139 82
Tubs	Number	2	1 50	4	3 70	6	5 20
Washboards	"	13	2 10			13	2 10
Wringers, (hand)	"	2	10 50			2	10 50
Total laundry			$773 70		$947 59		$1,721 29
FUEL.							
Charcoal	Bushels.	826	$73 46	1,506	$128 00	2,332	$201 46
Coal, anthracite	Tons			27¼	163 23	27¼	163 23
Coal, bituminous	"	2,979	2,946 05	2,781	2,617 72	5,760	5,563 77
Coke	Bushels.			100	6 00	100	6 00
Total fuel			$3,019 51		$2,914 95		$5,934 46
LIGHT.							
Alcohol	Gallons.	66½	$144 89	142½	$312 09	209	$456 98
Candles	Pounds.			40	4 40	40	4 40
Gasoline	Gallons.	7,905½	1,224 62	11,024	1,808 06	18,929½	3,032 68
Oil, kerosene	"	54½	11 58	153½	38 59	208	50 1
Oil, lard	"	99	59 40	95½	61 15	194½	120 55
Oil, sperm	"	48	30 24	184	91 65	232	121 89
Wick, lamp	Dozens.	10	65	12	75	22	1 40
Total light			$1,471 38		$2,316 69		$3,788 07

Itemized Statement—*Continued.*

Item.	Measure	1878-9. Am't.	1878-9. Cost.	1879-80. Am't.	1879-80. Cost.	1878 and 1880. Am't.	1878 and 1880. Cost.
MEDICINES AND MEDICAL SUPPLIES.							
Ale and beer			$76 70		$27 00		$103 70
Drugs, all sorts			1,168 79		848 45		2,017 24
Druggists' sundries			47 77		12 11		59 88
Instruments, med. and surg.			23 62	8	54 88		78 50
Liquors			246 12	103½	308 87		554 99
Means of restraint			142 76		65 62		208 38
Tobacco	Pounds.		444 56		434 80		879 36
Total medicines			$2,150 32		$1,751 73		$3,902 05
FREIGHT AND TRANSPORTATION.							
Case and cartage			$134 22		$160 74		$294 96
Drayage			2 00				2 00
Expenses of trustees			324 45		214 35		38 80
Expenses of legislature			49 31				49 31
Express charges			224 35		421 81		646 16
Freight, on coal			2,113 80		2,019 40		4,133 20
Freight, all other			1,524 06		1,955 53		3,479 59
Hauling			710 00		671 70		1,381 70
Livery bills					7 00		7 00
Returning fugitives			31 20		68 45		99 65
Transportation of inmates			109 55		107 13		216 68
Transportation of officers			209 26		201 12		410 38
Transportation of workmen			1 00				1 00
Total transportation			$5,433 20		$5,827 23		$11,260 43
POSTAGE, ETC.							
Newspaper wrappers			$1 00				$1 00
Postal cards	Number	950	9 50	1,950	$19 50		29 00
Stamps			364 47		190 00		554 47
Stamped envelopes					56 30		56 30
Telegraphing			71 20		47 08		118 28
Total postage			$446 17		$312 88		$759 05
BOOKS AND STATIONERY.							
Arm rests	Number	1	85			1	85
American annals	Copies.	4	$12 30			4	$12 30
Books, account	Number			13	$32 17	13	32 17
Books, blank	"	2	20 00			2	20 00
Books, medical	"		77 47				77 47
Books, day	"			1	4 80	1	4 80
Books, hymn	"	25	32 88			25	32 88
Books, library	"	3	16 82	49	62 74	52	79 56
Books, memorandum	"	6	1 75	60	14 50	66	16 25
Books, pass	"	72	4 10	48	1 00	120	5 10
Books, mem.			4 35				4 35
Books, school (S. S.)	Number	100	10 50			100	10 50
Books, scratch	"	72	6 30			72	6 30
Books, song	"			12	17 40	12	17 40
Books, cook	"	1	1 75			1	1 75
Cards		1,300	2 00	1,500	5 80	2,800	7 80
Envelopes	M.	¼	1 75			¼	1 75
Erasers, steel	Number	6	3 11	3	1 00	9	4 11
Erasers, rubber	"			144	90	144	90
Files, paper	"	100	2 00			100	2 00
Indexes	"	1	6 50			1	6 50
Ink	Quarts.	12	5 65	24	6 50	36	12 15
Ink, copying	Bottles.			1	30	1	30
Ink, red	"			6	1 75	6	1 75
Inkstands	Number	12	6 75	12	6 00	24	12 75
Journals	"	1	1 92	4	16 10	5	18 02
Journal Mental Science	Copies.	6	25 80	2	10 00	8	35 80

Itemized Statement—*Continued.*

Item.	Measure	1878-9. Am't.	1878-9. Cost.	1879-80. Am't.	1879-80. Cost.	1878 and 1880. Am't.	1878 and 1880. Cost.
Key-rings	Number			6	75	6	75
Ledgers	"	9	$21 85			9	$21 85
Magazines	"	4	12 65			4	12 65
Mucilage	Bottles..	24	3 10	48	$4 00	72	7 10
Newspapers, subscriptions to			37 12		75 45		112 57
Pads and blotters	"	432	1 50	144	60	576	2 10
Paper, blotting	Sheets..	12	60			12	60
Paper, legal cap	Reams..			1	3 24	1	3 24
Paper, manilla	Pounds..			20	1 70	20	1 70
Paper, note	Quires...	534	20 70			534	20 71
Paper, silver	Sheets..	12	2 94	12	38	24	3 32
Paper, tissue	Quires...	20	2 00	6	90	26	2 90
Paper, wrapping	Reams..	2	8 40			2	8 40
Pencils, lead	Dozens..	9	5 20	11	5 31	20	10 51
Pens, Fountain	Number	1	4 00			1	4 00
Pens, quill	Hundr'd	¼	75			¼	75
Pens, steel	Boxes ..	1	1 15		9 60		10 75
Pens, ruling	Number			3	1 44	3	1 44
Pens, rubber	"			48	1 17	48	1 17
Penholders	"			½	1 25	½	1 25
Railroad guides	"	1	20			1	20
Registers	"	1	8 75			1	8 75
Rubber stamps	"			2	1 50	2	1 50
Rubber pens	"	12	30			12	30
Rubber bands	Boxes ..	18	13 76			18	13 76
Rulers	"			3	1 13	3	1 13
Tablets	Number			431	16 55	431	19 55
Tags, shipping	"	350	50	1,000	4 25	1,350	4 75
Waste-paper baskets	"			2	1 50	2	1 50
Rubber stamp	"	1	4 00			1	4 00
Postal guide	"	1	1 00			1	1 00
Total books, etc			$395 03		$314 68		$709 71

PRINTING AND ADVERTISING.

Item.	Measure	1878-9. Am't.	1878-9. Cost.	1879-80. Am't.	1879-80. Cost.	1878 and 1880. Am't.	1878 and 1880. Cost.
Advertisements	Number			2	$5 50	2	$5 50
Affidavits	"			1,000	3 00	1,000	3 00
Applications	"			1,000	6 00	1,000	6 00
Bill-heads	"	900	$6 25	2,000	7 00	2,900	13 25
Blanks	"	9,500	59 85	11,100	51 00	20,600	110 85
Cards	"	50	2 00			50	2 00
Cards, postal	"	50	1 75	25	1 00	75	2 75
Circulars	"	100	2 00			100	2 00
Envelopes	"	14,000	38 00	11,000	38 00	25,000	76 00
Labels	M.	2	2 60	2	2 90	4	5 50
Letter-heads	"	3½	25 20	5	16 00	8½	41 20
Notices	Number			175	4 25	175	4 25
Orders, treasurer					13 00		13 00
Order-books	Number			1	9 00	1	9 00
Photographs	"	2	1 00			2	1 00
Programmes	"			450	2 50	450	2 50
Quarterly reports	"	75	30 00			75	30 00
Registers	"			2	18 00	2	18 00
Vouchers	"	200	7 00			200	7 00
Requisition blanks	C	18	13 00			18	13 00
Reports of condition			8 15				8 15
Pamphlets	Number	205	44 80			205	44 80
Dials				1,000	3 65	1,000	3 65
Total printing			$241 60		$180 80		$422 40

MUSIC AND AMUSEMENTS.

MUSIC.

Item.	Measure	1878-9. Am't.	1878-9. Cost.	1879-80. Am't.	1879-80. Cost.	1878 and 1880. Am't.	1878 and 1880. Cost.
Bows	Number	1	50			1	50
Cabinet organs	"			1	$45 00	1	$45 00
Pianos	"	1	$275 00	1	271 25	2	546 25
Piano stools	"	1	10 00	1	3 75	2	13 75
Repairs			8 00				8 00
Sheet-music			2 09		8 02		10 11

Itemized Statement—*Continued.*

Item.	Measure	1878-9. Am't.	1878-9. Cost.	1879-80. Am't.	1879-80. Cost.	1878 and 1880. Am't.	1878 and 1880. Cost.
Strings	Boxes ..	5	$5 50	38	$5 20		$10 70
Tamborines	Number			1	1 16	1	1 16
Tuning	Times ..	5	14 50	5	5 50	10	20 00
Violins	Number	1	4 70			1	4 70
Wire cord	Rolls....		1 37	39	7 40		8 77
Entertainments				3	55 00	3	55 00
AMUSEMENTS.							
Backgammon boards	Number	7	5 17			7	5 17
Billiard balls	"	4	11 50			4	11 50
Billiard chalk	Boxes ..	2	1 00			1	1 00
Billiard cloths	Number	1	20 00	1	10 63	2	30 63
Billiard cues	"			6	2 00	6	2 00
Billiard tips	Boxes ..			2	3 00	2	3 00
Bird baths	Number	12	90	24	3 00	36	3 90
Bir.¹ sced	Pounds..	60	4 60	95	6 30	155	10 90
Cages, bird	Number	11	18 74			11	18 74
Cards	Packs ..	120	32 85	72	12 00	192	44 85
Checkers	Boxes ..	12	1 20			12	1 20
Chess-boards	Number			6	6 00	6	6 00
Christmas gifts			10 13				10 13
Croquet balls	Number	1	2 00			1	2 00
Fish, gold	"	6	1 50			6	1 50
Flower-stands	"	2	13 00			2	13 00
Games	"	6	1 50	6	1 25	12	2 75
Magic lanterns, fixtures	Lot			1	15 75	1	15 75
Pictures	Number	14	50 00	32	68 23	46	118 23
Picture frames	"	12	19 50			12	19 50
Play-books	"	44	5 65	26	4 20	70	9 85
Slides for magic lantern	"			36	28 50	36	28 50
Theatrical properties	Lots	1	13 71	1	5 95	2	19 66
Lawn tennis			15 30				15 30
Dominoes		12	5 00			12	5 00
Hanging-baskets				36	13 00	36	13 00
Total amusements			$546 91		$590 09		$1,137 00

INSTRUMENTS AND APPARATUS.

MEDICAL AND SCIENTIFIC.

Item.	Measure	Am't.	Cost.	Am't.	Cost.	Am't.	Cost.
Bottles	Number	18	$2 65			18	$2 65
Druggists' sundries			75		7 50		8 25
Forceps, tooth	Number	4	7 65	2	3 50	6	11 15
Funnels	"	1	50			1	50
Lancets	"	1	63			1	63
Probangs	"	2	64			2	64
Rubber tubing	Feet	21	4 18	12	2 33	33	6 51
Syringes	Number	3	3 13	2	4 00	5	7 13
Syringes, hypodermic	"			1	2 00	1	2 00
Thermometers	"	8	6 47	2	5 00	10	11 47
Trusses	"			6	10 00	6	10 00
Evaporators	"	1	63			1	63
Roll-up pouch	"	1	1 50			1	1 50
Tongue depresser	"	1	1 50			1	1 50
Stomach tube	"	1	1 35			1	1 35
Bandages	"	2	1 70			2	1 70
Total apparatus			$33 28		$34 33		$67 61

HOUSEHOLD EXPENSES.

Item.	Measure	Am't.	Cost.	Am't.	Cost.	Am't.	Cost.
Ammunition (powder, shot)			$4 31				$4 31
Bags	Number			2	70	2	70
Bags, paper	"	12	60			12	60
Barrels	"	3	3 25			3	3 25
Barrel covers	"	12	1 25			12	1 25
Baskets	"	11	4 13	27	$18 50	38	22 63
Bath-brick	Boxes ..	7	5 25	16	10 75	23	16 00
Bed-bug exterminator			2 37				2 37
Blacking	Boxes ..	317	11 62	1,020	37 69	1,337	49 31

Itemized Statement—*Continued.*

Item.	Measure	1878-9. Am't.	1878-9. Cost.	1879-80. Am't.	1879-80. Cost.	1878 and 1880. Am't.	1878 and 1880. Cost
Bowls, wooden	Number			4	$1 66	4	$1 66
Brackets, wall	"			12	2 70	12	2 70
Brooms	"	534	$96 24	426	122 55	960	218 79
Brooms, whisk	"		1 25	12	4 18		5 43
Brushes, dust	"	60	24 50	60	23 75	120	48 25
Brushes, hair	"			24	7 93	24	7 93
Brushes, marking	"	6	43			6	43
Brushes, scrub	"	210	35 10	288	44 00	498	79 10
Brushes, shaving	"	12	1 25			12	1 25
Brushes, shoe	"	24	3 50	48	11 50	72	15 00
Brushes, tooth	"	12	2 00			12	2 00
Brushes, whitewash	"	1	1 50			1	1 50
Buckets, iron	"	1	1 00			1	1 00
Buckets, tin	"	7	3 00	6	4 05	13	7 05
Buckets, wooden	"	60	8 00	72	11 70	132	19 70
Carpet-stretchers	"			4	11 57	4	11 57
Chamois skins	"			3	75	3	75
Chopping-blocks	"	1	2 00			1	2 00
Chopping knives	"			7	1 25	7	1 25
Combs, course	"	96	7 59	300	32 02	396	39 61
Combs, fine	"	132	6 40	72	4 28	204	10 68
Disinfectant	Bottles		73 81	252	47 25		121 06
Dusters, feather	Number	272	114 67	48	16 00	290	130 67
Faucets	"	2	1 20			2	1 20
Flower pots, house	"	2,572	98 86			2,572	98 86
Flower-pot saucers	"			172	6 60	172	6 60
Fly traps	"	2	80			2	80
Fly fans	"			4	15 20	4	15 20
Fusees	Boxes	20	20 00	36	54 00	56	74 00
Insect powder	Pounds	15	9 65	3	1 75	18	11 40
Keys	Number	132	62 65		20 40		83 05
Lemon-squeezers	"			3	75	3	75
Matches	Boxes			2	4 77	2	4 77
Mops	Number	72	9 00	48	5 00	120	14 00
Mop sticks	"			72	7 50	72	7 50
Mop yarn	Pounds	40	27 00	144	18 00		45 00
Picture cord	Yards		70			40	70
Picture nails	Number	144	4 00			144	4 00
Refrigerators	"	1	19 50			1	19 50
Razors	"	6	5 00			6	5 00
Razor strops	"	4	4 00	7	4 00	11	8 00
Rolling pins	"			1	15	1	15
Rope	Pounds	89	9 16	20	2 80	109	11 96
Rubber rings (fruit can)	Dozens	2	50			2	50
Sapolio	Cakes	401	77 50	3,029	217 50	3,430	295 00
Saw-dust		30	45 00			30	45 00
Scales	Number	2	16 00			2	16 00
Scale-weights	Setts	3	3 60			3	3 60
Scissors	Pairs	2	1 85	13	9 42	15	11 27
Sealing wax, for cans	Pounds			20	1 00	20	1 00
Shears	Pairs			6	4 50	6	4 50
Sieves	Number	1	83	2	3 50	3	4 33
Silver polish	Boxes	144	6 50	288	17 00	432	23 50
Soap, castile	Pounds	2	50			2	50
Soap, shaving	Boxes			4	3 75	4	3 75
Soap, toilet	"	180	20 07	164	16 36	344	36 43
Sponges	Number			15	26 25		52 45
Step-ladders	"	3	7 50			3	7 50
Tacks	Papers		9 78	456	24 94		34 72
Tack hammers	Number			24	2 45	24	2 45
Tobacco-pipes	Boxes	1	1 65			1	1 65
Traps, mouse	Number	6	1 50	12	75	18	2 25
Traps, rat	"			15	7 88	15	7 88
Tripoli	Pounds			24	14 40	24	14 40
Twine	"	2	95	3	85	5	1 80
Zinc, boards	Number	2	4 00			2	4 00
Lamp brush	Number	1	25			1	25
Rotten stone	Pounds	25	1 75			25	1 75
Egg twirler	Number	1	1 50			1	1 50
Disinfectant	Gallons				111 60		111 60
Carpet sweepers (patent)	Number			6	12 00	6	12 00
Chalk					60		60
China eggs				15	75	15	75
Total household expenses			$913 97		$1,031 20		$1,945 17

Itemized Statement—*Continued.*

Item.	Measure	1878-9. Am't.	1878-9. Cost.	1879-80. Am't.	1879-80. Cost.	1878 and 1880. Am't.	1878 and 1880. Cost.
FURNITURE.							
CLASS 1—Manufactured.							
Bedsteads	Number	25	$187 50	46	$358 00	71	$545 50
Bookcases	"	1	7 00			1	7 00
Bracket	"	1	16 00			1	16 00
Bureaus	"	14	124 00	9	72 25	23	196 25
Chairs	"	68	67 85	53	203 40	121	271 25
Clocks	"	3	19 62	5	5 60	8	25 22
Cribs	"			1	30 00	1	30 00
Desks, office	"			3	33 00	3	33 00
Looking-glasses	"	12	6 50	6	2 13	18	8 63
Settees, sofas	"	6	33 90			6	33 90
Sewing machines	"	1	30 00			1	30 00
Stands	"	2	2 00			2	2 00
Tables	"	7	53 50	1	8 00	8	61 50
Towel-racks	"	1	1 50			1	1 50
Upholstering, staples	Papers			2	1 00	2	1 00
Vases	Number	2	6 00			2	6 00
Book-rack	"	1	1 60			1	1 60
Wash-stands	"	1	4 00	7	14 00	8	18 00
CLASS 2—Doors and Windows.							
Awnings	Yards	54	10 90	56½	10 59	110½	21 49
Carpets	"	85¼	55 74	454¼	455 90	539½	511 64
Carpet paper	Pounds	179	17 80	21	1 68	200	19 48
Curtain goods	Yards		1 13		47 44		48 57
Curtain fixtures	Sets			30	24 27	30	24 27
Linoleum	Yards	133½	130 67			133½	130 67
Mats, floor	Number	7	10 53	2	2 21	9	12 74
Matting	Yards	66	54 98			66	54 98
Oil cloth	"	4	12 87	38	18 84	42	31 71
Rugs	Number	12	52 82	2	10 88	14	63 70
Shades	Lots		29 07	4	58 47		87 54
Stair carpet	Yards	90	92 54			90	92 54
Tassels	Number			4	1 11	4	1 11
Repairing clocks				2	2 50	2	2 50
Repairing sewing machines				4	12 75	4	12 75
CLASS 3—For Beds.							
Hair	Pounds			50	17 32	50	17 32
Mattresses, wire	"	4	28 50			4	28 50
Mattress needles	"	6	1 75			6	1 75
Mattress twine	Pounds	14½	11 18	9	5 40	23½	16 58
Renovating and repairs			24 25				24 25
Mattress-makers' findings			1 31				1 31
CLASS 4—Glass, Queensware and Cutlery.							
Basins	Number			12	4 50	12	4 50
Bowls	"	60	8 00		20 90		28 90
Bowls, sugar	"	24	9 00			24	9 00
Carvers	Sets	12	25 00			12	25 00
Castors	Number			12	30 00	12	30 00
Chambers	"	216	151 20	134	119 48	350	270 68
Chamber, toilet sets	"	3	13 00			3	13 00
Cruets	"	72	11 25	144	36 00	216	47 25
Cups	"	420	51 63	84	7 13	504	58 76
Cuspidors	"	36	33 66	9	6 75	45	40 41
Dishes, butter	"	36	1 20	24	10 80	60	12 00
Dishes, glass	"	6	38	3	1 50	9*	1 88
Dishes, sauce	"	12	55	6	1 25	18	1 80
Dishes, vegetable	"	24	16 00	134	33 62	158	49 62
Ewers	"	12	8 40	12	8 10	24	16 50
Fruit jars	"	1	15	2	70	3	85
Glasses, jelly	"	72	6 20	60	5 00	132	11 20
Glasses, medicine	"	144	3 00			144	3 00
Globes	"	1	1 75			1	1 75
Goblets	"	414	33 50	36	5 25	450	38 75
Jars	"	2	92			2	92
Jugs	"	27	5 15	25	6 15	52	11 30

Itemized Statement—*Continued.*

Item.	Measure	1878-79. Am't.	1878-79. Cost.	1879-80. Am't.	1879-80. Cost.	1878 and 1880. Am't.	1878 and 1880. Cost.
Knives, butcher	Number			2	$2 11	2	$2 11
Knives, butter	"	12	$5 00			12	5 00
Knives and forks	"	240	32 75			240	32 75
Ladles	"	24	11 00			24	11 00
Lamps	"	11	39 17	6	1 75	17	40 92
Lamp chimneys	"	78	4 40	144	8 70	222	13 10
Lamp burners	"	30	4 94	24	3 40	54	8 34
Lanterns	"	1	85	2	2 25	3	3 10
Lantern globes	"	18	3 15			18	3 15
Glass chimneys	Dozen	480	22 52	650	32 50	1,130	55 02
Pitchers	"		84 20	12	11 25		95 45
Pitchers, cream	"	24	4 50			24	4 50
Pitchers, molasses	"	46	20 13	14	6 13	60	26 26
Plates, breakfast	"			144	11 16	144	11 16
Plates, dinner	"	144	15 24	144	15 24	288	30 48
Plates, tea	"	144	11 16	72	10 80	216	21 96
Plates, pie	"			72	3 25	72	3 25
Plates, soup	"	48	4 91	396	36 06	444	40 97
Platters	"	48	6 38			48	6 38
Saucers	"	144	10 80	12	65	156	11 45
Soap dishes	"	24	2 20	12	5 00	36	7 20
Spittoons	"			1	1 15	1	1 15
Spoons, mustard	"			72	2 40	72	2 40
Spoons, tea	"	144	4 00			144	4 00
Steels	"			5	2 70	5	2 70
Strainers	"	1	28	1	2 50	2	2 78
Forks	"	48	19 40			48	19 40
Stoppers, glass	"	36	2 25			36	2 25
Tumblers	"	72	1 50	72	3 60	144	5 10
Tureens	"	1	4 00			1	4 00
CLASS 5—*Tin, Iron, Sundries.*							
Bells, dinner	Number	7	5 85			7	5 85
Bread trays	"			6	2 25	6	2 25
Broilers	"			12	4 50	12	4 50
Buckets	"	26	22 05	15	17 45	41	39 50
Cabbage-cutters	"	1	1 50			1	1 50
Caddies	"			51	9 00	51	9 00
Cake-turners	"	24	1 76			24	1 76
Cans, fruit	"			360	39 00	360	39 00
Cans, milk	"	24	24 00			24	24 00
Cans, food	"			53	40 00	53	40 00
Cleavers	"			1	1 42	1	1 42
Coal-scuttles (hods)	"	6	7 62	6	8 00	12	15 62
Coffee mills	"	1	67			1	67
Coffee pots	"			8	27 00	8	27 00
Covers, tin	"	16	32 00			16	32 00
Cullenders	"			2	6 60	2	6 60
Cups, tin	"			24	1 50	24	1 50
Dippers	"	24	2 20	12	4 02	36	6 22
Dredge-boxes	"	6	30			6	30
Dustpans	"	24	4 00	24	4 50	48	8 50
Egg beaters	"	1	65	12	60	13	1 25
Elbows	"			4	1 80	4	1 80
Fire shovels	"	1	70			1	70
Flesh-forks	"	24	4 80		1 25	6	6 05
Funnels	"	2	40			2	40
Gas stoves	"	1	6 30			1	6 30
Gem irons	"	6	1 50			6	1 50
Graniteware	"	40	30 35	6	10 00	46	40 35
Ice-cream freezers	"	1	5 00			1	5 00
Kettles, iron	"	1	45	1	120 00	2	120 45
Meat saws	"			1	1 42	1	1 42
Napkin rings	"	12	1 25			12	1 25
Pans, dish	"			6	8 00	6	8 00
Pans, drip	"	12	15 00			12	15 00
Pans, iron	"	80	86 96			80	86 96
Pans, meat sauce	"			28	13 70	28	13 70
Pans, milk	"	2	1 15	6	75	8	1 90
Pans, pie	"			144	5 00	144	5 00
Pokers	"	1	70			1	70
Pots, iron	"			1	65	1	65
Pots, coffee	"	1	2 00			1	2 00
Pots, tea	"	6	5 60			6	5 60

Itemized Statement—*Continued.*

Item.	Measure	1878-9. Am't.	1878-9. Cost.	1879-80. Am't.	1879-80. Cost.	1878 and 1880. Am't.	1878 and 1880. Cost.
Repairs on tinware			$4 30		$70 00		$74 30
Scoops	Number	9	10 50			9	10 50
Skimmers	"	1	20	1	49	2	69
Spiders	"			6	1 80	6	1 80
Sprinklers	"	7	2 95	45	14 85	52	17 80
Stoves	"			2	21 25	2	21 25
Stove-pipe	"				8 80		8 80
Thimbles	"			4	60	4	60
Potatoe mashers	"			1	50	1	50
Tongs	"	1	70			1	70
Trays	"	3	3 00	8	7 60	11	10 60
Waffle-irons	"	1	60			1	60
Water-coolers	"	3	8 25			3	8 25
Wash-boilers	"			2	4 25	2	4 25
Total furniture			$2,086 94		$2,315 02		$4,401 96

BUILDING, IMPROVEMENTS AND REPAIRS.

BRICKWORK AND PLASTERING.

Brick, asphaltum	M	3½	$184 98	30	$900 00	33½	$1,084 98
Cement	Barrels	44	125 40	36	83 00	80	208 40
Fire-brick	Number	300	6 75	300	3 00	600	9 75
Fire-clay	Barrels			2	2 50	2	2 50
Lath	M	4	6 00		5 50	6	11 50
Lime	Bushels	337	83 82	260	57 20	597	141 02
Sand	Loads	106	216 56	94	175 60	200	392 16
Stone, dimension	Yards		69 75				69 75
Stone, rubble	"	14	7 00			14	7 00
Stucco	Barrels	10	33 15	4	12 50	14	45 65

CARPENTER WORK.

Doors	Number			8	22 40	8	22 40
Lumber, dressed	Feet	27,600	468 40	14,700	492 55	42,300	960 95
Lumber, flooring	"	1,000	20 00			1,000	20 00
Lumber, hard	"	608	21 84		90 38		112 22
Lumber, pine	"			11,494	147 33	11,494	147 33
Lumber, poplar	"	88,883	270 94	11,845	157 48	100,728	428 42
Lumber, yellow pine	"				155 83		155 83
Moulding, beads	"	150	9 45	891	34 73	1,041	44 18
Sashes	Number	43	43 00	4	4 95	47	47 95

Hardware.

Bells	Number	1	2 75			1	2 75
Bolts	"	743	21 26	315	29 25	1,058	50 51
Bolts, brass	"			2	20	2	20
Butts	Pairs	48	4 60		28 34		32 94
Butts, brass	"	12	18			12	18
Chain, brass	Number	156	17 21		7 80		25 01
Clamps	"			14	4 75	14	4 75
Draw-pulls	"			64	3 45	64	3 45
Escutcheons	"			26	13 00	26	13 00
Glue	Pounds	211	29 15			211	29 15
Glue, white	"	97	34 26			97	34 26
Hasps	Number	24	2 05			24	2 05
Hinges	Pairs	137	37 25	41	14 59	178	51 84
Hooks	Number	553	8 95	1,642	25 08	2,195	34 03
Hooks, clothes	"	432	4 00			432	4 00
Knobs	"	24	6 72	24	9 50	48	16 22
Locks	"	50	39 73	117	64 29	167	104 02
Nails	Pounds	412	16 09	2,954	127 22	3,366	143 31
Nails, finishing	Papers	11	1 06	59	9 89	70	10 95
Pulleys	Number	24	2 80			24	2 80
Rings	"			12	1 25	12	1 25
Sash-cord	Pounds			48	29 60	48	29 60
Sash-weights	Number			60	15 60	60	15 60
Screws	Gross	96½	43 81	36¾	21 24	133	65 05
Screws, bench	Number			3	1 88	3	1 88
Springs	"	16	29 31	49	7 15	65	36 46
Staples	One lot	1	75			1	75

—4

Itemized Statement—*Continued.*

Item.	Measure	1878-9. Am't.	Cost.	1879-80. Am't.	Cost.	1878 and 1880. Am't.	Cost.
Transom lifts	Number	6	$8 67			6	$8 67
Washers	Pounds			2½	37	2½	37
Wire cloth	Feet	1,294	140 83	506	$131 74	1,800	272 57
Window guards	Number	5	25 00			5	25 00
PAINTING AND GLAZING.							
Burnt umber	Pounds	190	34 08	1	15	191	34 23
Glass	Boxes	1	3 67	17	87 95	18	91 62
Glass	Lights	6	3 85	20	22 45	26	26 30
Gold bronze	Pounds	½	2 00			½	2 00
Gum schellac	"	2	1 40			2	1 40
Japan	Gallons	20	15 50	5	4 25	25	19 75
Lampblack	Pounds	20	3 60	1	20	21	3 80
Oehre, red	"	3	15			3	15
Oils	Gallons	335	234 25	642	163 72	977	397 97
Paints, chemical	Pounds			200	25 20	200	25 20
Paints, lead	"	9,687	668 96	3,817	312 28	13,504	981 24
Paints, mineral	"	703	33 68			703	33 68
Preservative	Gallons			103½	163 05	103½	163 05
Putty	Pounds	113	4 11	281½	9 94	394½	14 05
Turpentine	Gallons	70½	23 65	115¾	59 65	186¼	83 30
Varnish	"	27	23 50	65⅝	106 95	92⅝	130 45
Paints, miscellaneous			50 45	1	25		50 70
ROOFING.							
Bends	Number			3	2 64	3	2 64
Down-spouts	Feet			114	23 40	114	23 40
Elbows	Number	4	1 19	23	4 25	27	5 44
Gutters	Feet			257	47 00	257	47 00
Shingles	M			58½	151 95	58½	151 95
Slate	Squares	28	161 00			28	161 00
Repairs					2 75		2 75
METALS.							
Copper	Pounds	182	50 96			182	50 96
Iron	"	35	1 33	141	8 30	176	9 63
Iron, galvanized	"	347	31 23	162	25 11	509	56 34
Lead, sheet	"			2,595	217 10	2,595	217 10
Tin	Boxes	1	9 50	4	44 00	5	53 50
Zinc	Pounds	85	6 80			85	6 80
IRON PIPE.							
Caps	Number	30	76			30	76
Cocks	"	46	91 58	39	102 42	85	194 00
Connexions	"	69	7 43	22	1 37	91	8 80
Ells	"	217	16 42	187	8 81	404	25 23
Fittings	Lots	2	15 96	2	21 30	4	37 26
Goosenecks	Number			3	90	3	90
Hangers	"	20	1 68	15	5 63	35	7 31
Joints	"			24	5 28	24	5 28
Nipples	"	233	15 85			233	15 85
Offsets	"	1	84			1	84
Pipe, brass	Pounds			9¾	5 20	9¾	5 20
Pipe, gas	"	1	1 20			1	1 20
Pipe, iron	Feet	925	53 27	1,659	140 99	2,584	194 26
Reducers	Number	12	35			12	35
Tees	"	107	11 52	829			19 81
Traps	"	8	25 55			8	25 55
Valves	"	9	9 77	22	36 29	31	46 06
Washers	"			200	1 70	200	1 70
Cess-pools	"	2	4 88			2	4 88
GAS-FITTING.							
Brackets	Number	1	9 50	2	2 20	3	11 70
Globes	"			12	8 00	12	8 00
Pendants	"			5	22 25	5	22 25
Reflectors	"	72	13 50	31	5 82	103	19 32
Cutter wheels	"	2	34			2	34
Fitters, cement	Pounds	2	50			2	50
Shades	Number			10	4 17	10	4 17

Itemized Statement—*Continued.*

Item.	Measure	1878-9. Am't.	1878-9. Cost.	1879 80. Am't.	1879 80. Cost.	1878 and 1880. Am't.	1878 and 1880. Cost.
Torches	Number			12	$2 75	12	$2 75
Shade holders	"	1		12	3 50	12	3 50
PLUMBING.							
Bibbs	Number	11	$31 45	18	18 60	29	50 05
Clamps	"			12	2 70	12	2 70
Force-cups	"			12	6 00	12	6 00
Force-pump	"	1	18 85			1	18 85
Plugs	"	110	3 45	34	4 52	144	7 97
Sinks	"			1	4 60	1	4 60
Urinals	"	1	2 31			1	2 31
Wash-stands	"	1	5 25			1	5 25
Water closets	"	1	26 70			1	26 70
Slop-tanks	"			3	141 00	3	141 00
Filters	"			13	26 00	13	26 00
OTHER IRON WORK.							
Bolts	Number	56	13 20			56	13 20
Bushings	"	99	5 08	24	69	123	5 77
Car wheels	"	6	10 80			6	10 80
Castings	"	299	36 55	1,090	30 91	1,389	67 46
Castings, brass	"	4	2 60			4	2 60
Cotton waste	Pounds.	100	10 50			100	10 50
Crocus	"	30	3 50			30	3 50
Gauges	Number	1	8 00	2	4 36	3	12 36
Grate bars	Pounds.	4,350	108 75			4,350	108 75
Hog chains	"			110	12 00	110	12 00
Packing	"			55½	20 95	55½	20 95
Registers	"			5	24 00	5	24 00
Rivets	"	36	3 68			30	3 68
Steam traps	Number			1	28 00	1	28 00
Steam whistles	"	1	3 15			1	3 15
Rollers	"	18	22 50			18	22 50
EXCAVATION AND SEWERAGE.							
Drain tile	Number			72	10 80	72	10 80
Sewer pipe	"	1,482	1,086 02	261	77 25	1,743	1,160 27
Sewer joints	"	26	25 75	14	8 96	40	34 71
WORK DONE BY JOB OR CONTRACT.							
Building			1,575 00		1,017 38		2,592 38
Bricklaying			304 85				304 85
Monthly estimates (contract).	W.sup'ly		795 00				795 00
Repairs			4 00		88 50		92 50
MISCELLANEOUS.							
Emery	Pounds.	5	50			5	50
Emery paper	Quires.			16½	5 01	16½	5 01
Hose (suction and con.)	Feet		83 94	800	165 00		248 94
Sand paper	Quires.	6	1 32	36	7 63	42	8 95
Spanners	Number	2	1 50	4	1 00	6	2 50
Tiling	Feet		494 64	315	78 75		573 39
Ventilators	Number			3	33 75	3	33 75
Wall paper	Bolts	7	8 75			7	8 75
Wall paper border	"	8	42 43			8	42 43
Wire	"	5	2 25	61½	20 29	66½	22 54
Wire rope	Feet	180	14 40			180	14 40
Table legs	Number	24	4 80	8	1 20	32	6 00
Brass letters	"	156	15 60			156	15 60
Total repairs			$8,466 30		$6,597 40		$15,063 70
TOOLS.							
Bits	Sets			1	$8 00	1	$8 00
Brushes, paint	Number	66	$75 57	8	13 35	74	88 92
Cutters	"	1	1 50	2	1 00	3	2 50
Files	"	23	3 50	132	24 35	155	27 84

52

Itemized Statement—*Continued.*

Item.	Measure	1878 9. Am't.	Cost.	1879-80. Am't.	Cost.	1878 and 1880. Am't.	Cost.
Graining combs	Sets	1	$1 60			1	$1 60
Hammers	Number			12	$5 63	12	5 63
Hatchets	"	6	4 00	6	3 50	12	7 50
Levels	"	1	1 85			1	1 85
Lines, chalk	"	3	30	12	2 25	15	2 55
Oil cans	"			9	16 05	9	16 05
Oilers	"	6	1 38	6	1 80	12	3 18
Sash tools	"	12	2 10	11	2 34	23	4 44
Saws	"	2	2 56	9	11 19	13	13 75
Scrapers	"	1	33			1	33
Screw drivers	"			12	3 00	12	3 00
Shovels	"			12	10 50	12	10 50
Tongs	"	1	1 30	3	3 30	4	4 60
Trowels	"	1	1 50			1	1 50
Vises	"			1	5 00	1	5 00
Wrenches, monkey	"	3	2 55	1	1 50	4	4 05
Scoops	"			6	5 75	6	5 75
Total tools			**$100 04**		**$118 50**		**$218 54**

MACHINERY, ETC.

Item.	Measure	1878 9. Am't.	Cost.	1879-80. Am't.	Cost.	1878 and 1880. Am't.	Cost.
Belting, leather	Feet			19½	$9 63	19½	$9 63
Belting, rubber	Number	60	$12 02			60	12 02
Cranks	"	12	50			12	50
Engines	"	1	300 00			1	300 00
Fans	"	1	170 00			1	170 00
Grindstones	"			2	5 50	2	5 50
Mortising machines	"			1	47 50	1	47 50
Pumps, Worthington	"	2	3,180 00			2	3,180 00
Ranges	"			1	488 00	1	488 00
Saws, band, (scroll)	"			1	24 50	1	24 50
Saws, blades, (scroll)	"			240	13 10	240	13 10
Steamers	"	4	173 33	1	40 00	5	213 33
Washing machines	"	3	775 00			3	775 00
Repairs to machinery	"		352 89		103 49		456 38
Hose carriage	"	1	65 00			1	65 00
Brick machine	"	1	30 00			1	30 00
Lace leather	Side	1	3 00			1	3 00
Total machinery, etc			**$5,061 74**		**$731 72**		**$5,793 46**

FARM, GARDEN, STOCK AND GROUNDS.

FEED.

Item.	Measure	1878 9. Am't.	Cost.	1879-80. Am't.	Cost.	1878 and 1880. Am't.	Cost.
Feed	Pounds	60,950	$319 05	78,543	$393 70	139,493	$712 75
Oats	Bushels			25	12 00	25	12 00
Straw	Tons		21 50		29 00		50 50

LIVE STOCK.

Item.	Measure	1878 9. Am't.	Cost.	1879-80. Am't.	Cost.	1878 and 1880. Am't.	Cost.
Horses	Number	3	330 00	1	80 00	4	410 00

VEHICLES.

Item.	Measure	1878 9. Am't.	Cost.	1879-80. Am't.	Cost.	1878 and 1880. Am't.	Cost.
Wagons	Number	1	120 00	1	72 00	2	192 00

HARNESS.

Item.	Measure	1878 9. Am't.	Cost.	1879-80. Am't.	Cost.	1878 and 1880. Am't.	Cost.
Bits	Number	2	40			2	40
Collar pads	"	4	2 00			4	2 00
Collars	"			1	2 00	1	2 00
Grease, axle	Boxes				16 50		16 50
Halters	Number	3	4 50	2	60	5	5 10
Harness	"	3	48 50			3	48 50
Hitching straps	"	2	1 25	2	1 00	4	2 25
Horse weights	"	2	1 56			2	1 56
Oil, castor	Gallons	1	1 20	6	3 35	7	4 55
Oil, neats foot	"	3	3 75	2	2 00	5	5 75
Reins	Pairs	1	2 25	2	6 00	3	8 25
Robes	Number	1	8 50			1	8 50
Straps	"	6	45			6	45
Saddles	"			1	15 00	1	15 00
Whips	"	7	6 30	3	2 85	10	9 15

Itemized Statement—*Continued.*

Item.	Measure	1878-9. Am't.	1878-9. Cost.	1879-80. Am't.	1879-80. Cost.	1878 and 1880. Am't.	1878 and 1880. Cost.
AGRICULTURAL IMPLEMENTS.							
Grain drills	Number	2	$66 10			2	$66 10
Hand carts	"	2	16 00			2	16 00
Horse-rakes	"	1	30 00			1	30 00
Lawn mowers	"	1	16 80	1	$18 75	2	35 55
Mowers	"	1	88 20			1	88 20
Boilers	"			1	21 00	1	21 00
Wheelbarrows	"	8	18 75			8	18 75
Slop cart	"			1	75 00	1	75 00
FARM AND GARDEN TOOLS, ETC.							
Axes	Number	3	3 00			3	3 00
Axe handles	"	24	4 90	30	3 05	54	7 95
Forks, hay	"			6	2 60	6	2 60
Garden syringe	"	1	7 00			1	7 00
Hoes	"	12	3 90	12	4 20	24	8 10
Measures	"	1	75	1	40	2	1 15
Picks	"			6	5 00	6	5 00
Post-hole diggers	"			1	3 50	1	3 50
Pumps	"	1	2 20			1	2 20
Rakes	"	3	1 65	3	1 75	6	3 40
Sacks	"	1	20			1	20
Scythes	"	8	6 65	6	10 20	14	16 85
Scythe-stones, (whetstones)	"	12	1 00	3	25	15	1 25
Shovels	"			24	22 83	24	22 83
Spades	"			6	6 50	6	6 50
Scoops	"			6	5 63	6	5 63
SEEDS, ETC.							
Timothy seed	Bushels			3	9 75	3	9 75
Clover	"			3	17 25	3	17 25
Onion sets	"			27	31 08	27	31 08
Scions	"			71	20 00	71	20 00
Seeds	"		133 74		77 86		211 60
Seed corn	"			1½	13 50	1½	13 50
Seed potatoes	"			56	48 50	56	48 50
Shrubs and shrubbery	Number	200	5 00	283	17 50	483	22 50
Plants, Str. B	"	5,000	18 00			5,000	18 00
Grass seed	Bushels			10	25 00	10	25 00
ROADS AND FENCING.							
Gravel	Loads			1,578	144 94	1,578	144 94
Posts	Number			12	9 50	12	9 50
FARM REPAIRS.							
Axle washers	Number		50				50
Blacksmithing and shoeing			197 80		189 53		387 33
Carriage bolts	Number		1 70	4	40		2 10
Repairs to carriages			13 50		53 65		67 15
Repairs to harness			9 35		26 75		36 10
Repairs to tools and implements			8 45		42 75		51 20
Repairs to wagons			30 90		45 29		76 19
Total farm, etc			$1,557 25		$1,589 91		$3,147 16
LEGAL EXPENSES.							
Attorney's services			$327 90				$327 90
Notary public			28 70		$33 15		61 85
Police					3 65		3 65
Total			$356 60		$36 80		$393 40

Itemized Statement—*Continued.*

Item.	Measure.	1878-9. Am't.	Cost.	1879-80. Am't.	Cost.	1878 and 1880. Am't.	Cost.
SHOP EXPENSES.							
SHOE SHOP.							
Pincers	Pairs			1	50	1	50
Awls, pegging	Number		40	36	$5 70		$6 10
Measures, shoe	"			2	70	2	70
Shoe-stretchers	"			2	2 50	2	2 50
Boot-floats	"			2	1 38	2	1 38
Bristles	Ounces		65	5	2 75	6	3 40
Hammers	Number			1	29	1	29
Knives	"			1	08	1	08
Lasts	"			4	1 60	4	1 60
Leather, upper	Sides	2	$18 00	2	9 43	4	27 43
Leather, sole	Pounds.	67	26 14	22½	16 46	89½	42 60
Pegs	Quarts				1 45		1 45
Rasps	Number			1	23	1	23
Shoe-nails	Papers				81		81
Shoe-thread	Balls		2 85		5 00		7 85
Wax			29		62		91
CHAIR SHOP.							
Chair-bottoms	Number			120	32 84	120	32 84
PRINTING OFFICE.							
Paper	Reams	1½	1 25			1½	1 25
Printing ink	Cans	1	75		4 55		5 30
Type	Pounds		22 51				22 51
Type cases	Number	5	2 70			5	2 70
Type rollers	"	1	30			1	30
Rules					2 45		2 45
Walnut brackets				48	96	48	96
Total shop			**$75 84**		**$90 30**		**$166 14**
BURIAL EXPENSES.							
Coffins and boxes		21	$131 00	14	$94 00	35	$225 00
EXPENSES NOT CLASS-IFIED.							
Money refunded					$30 15		$30 15
Sewing machine repairs			$3 54				3 54
Can-openers			5 00				5 00
Leather			50				50
Rat-killers		416	4 16	694	6 94		11 10
Oil-stone				1	6 35	1	6 35
Curtain-pole				1	17 82	1	17 82
Nozzles					50		50
Whale-oil soap	Boxes			1	20	1	20
Bull ring				1	60	1	60
Total			**$13 20**		**$62 56**		**$75 76**

TREASURER'S REPORT.

R. B. STINSON, *Treasurer, in account with the Illinois Southern Hospital for the Insane, for the year ending September 30, 1879.*

ORDINARY EXPENSE.

Dr.

1878.				
October 1.....	To balance on hand...............................		$18,848 95	
November 8..	Amount from state treasurer.....................		21,250 00	
1879.				
January 20...	Amount from state treasurer.....................		21,250 00	
April 21.......	. Amount from state treasurer.....................		21,250 00	
July 21........	Amount from state treasurer.....................		21,250 00	
	Sundry deposits by Dr. H. Wardner, superintend't		9,172 52	
	Sundry deposits by John E. Detrich, president board trustees, on judgment against ex-treasurer and his sureties...............................		2,154 42	

Cr.

1879.				
September 30.	By superintendent's orders paid.................			$87,214 38
	Balance			27,961 51
			$115,175 89	$115,175 89
October 1.....	To balance		$27,961 51	

IMPROVEMENTS AND REPAIRS.

Dr.

1878.				
October 27....	To amount from state treasurer.................		$1,649 80	
1879.				
January 15...	Amount from state treasurer....................		1,967 84	
April 24.......	Amount from state treasurer....................		816 04	
August 13.....	Amount from state treasurer....................		1,702 11	
September 30.	To balance, overdraft...........................		2,022 45	

Cr.

1878.				
October 1.....	By balance, overdraft..........................			$1,278 87
1879.				
September 30.	By superintendent's orders paid................			6,879 37
			$8,158 24	$8,158 24
October 1.....	By balance....................................			$2,022 45

IMPROVEMENT OF GROUNDS.

Dr.

1878.				
October 27....	To amount from state treasurer.................		$279 98	
1879.				
January 20 ...	Amount from state treasurer....................		186 62	
April 24.......	Amount from state treasurer....................		5 00	
August 13.....	Amount from state treasurer....................		160 50	
	To balance		108 50	

Cr.

1878.				
October 1.....	By balance, overdraft..........................			$208 74
1879.				
September 30.	By superintendent's orders paid................			531 86
			$740 60	$740 60
October 1.....	By balance, overdraft..........................			$108 50

Treasurer's Report—*Continued*.

ROAD FROM ANNA.
Dr.

1878.
October 1..... To balance on hand $3 45

Cr.

December 28. By superintendent's order paid....................... | .... | $3 45
| | $3 45 | $3 45

CARPENTER SHOP.
Dr.

1879.
August 13..... To amount from state treasurer..................... $209 39

Cr.

1879.
September 30. By superintendent's orders paid.................... | .... | $209 39
| | $209 39 | $209 39

FRAME BARN.
Dr.

1878.
October 27.... To amount from state treasurer..................... $42 08

Cr.

1879.
October 1..... By balance, overdraft .... | .... | $42 08
| | $42 08 | $42 08

FIRE PUMP.
Dr.

1879.
January 20 ... To amount from state treasurer..................... $1,800 00

Cr.

1879.
September 30. By superintendent's orders paid.................... | .... | $1,800 00
| | $1,800 00 | $1,800 00

ROTARY OVEN.
Dr.

1878.
August 24..... To amount from state treasurer..................... $99 94

Cr.

1879.
September 30. By superintendent's orders paid.................... | .... | $99 94
| | $99 94 | $99 94

NEW KITCHEN.
Dr.

1879.
August 13..... To amount from state treasurer.................. $8 00
October 30.... Balance, overdraft................................ 357 03

Cr.

1879.
September 30. By superintendent's orders paid.................... | .... | $365 03
| | $365 03 | $365 03

October 1..... By balance................................ | .... | $357 03

Treasurer's Report—*Continued.*

	WATER SUPPLY.		
	Dr.		
1879.			
August 13.....	To amount from state treasurer......................	$795 00	
September 30.	Balance, overdraft...........................	1,020 68	
	Cr.		
1879.			
September 30.	By superintendent's orders paid	....	$1,815 68
		$1,815 68	$1,815 68
October 1.....	By balance ....'...........	....	$1,020 68

	REMOVAL OF BARN.		
	Dr.		
1879.			
September 30.	To balance, overdraft.........................	$175 00	
	Cr.		
1879.			
September 30.	By superintendent's orders paid	....	$175 00
		$175 00	$175 00
October 1.....	By balance................................	....	$175 00

	EXTENDING SEWER.		
	Dr.		
1879.			
September 30.	To balance, overdraft.........................	$921 22	
	Cr.		
1879.			
September 30.	By superintendent's orders paid...............	....	$921 22
		$921 22	$921 22
October 1.....	By balance................................	....	$921 22

	RECAPITULATION.		
	Dr.		
1879.			
October 1.....	To balance, ordinary expense................	$27,961 61	
	Cr.		
September 30.	By overdraft, improvement and repairs................	....	$2,022 45
	Overdraft, improvement of grounds	....	108 50
	Overdraft, new kitchen.......................	....	357 03
	Overdraft, water supply.....................	....	1,020 68
	Overdraft, moving old barn...................	....	175 00
	Overdraft, extending sewer...................	....	921 22
	Total balance on hand................	....	23,356 63
		$27,961 51	$27,961 51

TREASURER'S REPORT.

R. B. STINSON, *Treasurer, in account with the Illinois Southern Hospital for the Insane, for the year ending September 30, 1880.*

ORDINARY EXPENSE.
Dr.

1879.			
October 1.....	To balance on hand...........................	$27,961 51	
November 19.	Amount from state treasurer................	16,500 00	
1880.			
January 30....	Amount from state treasurer................	16,500 00	
April 23......	Amount from state treasurer................	16,500 00	
July 12........	Amount from state treasurer................	16,500 00	
	Sundry deposits by Dr. H. Wardner, Supt........	8,078 74	
	Sundry deposits by John F. Detrich, president of trustees, for redemption of land sold on judgment against ex-treasurer and his sureties....	814 80	

Cr.

1880.			
September 30.	By superintendent's orders paid................		$89,287 75
	Balance		13,567 30
		$102,855 05	$102,855 05
1880.			
October 1.....	To balance..................................	$13,567 30	

IMPROVEMENT AND REPAIRS.
Dr.

1879.			
October 31....	To amount from state treasurer............	$2,582 44	
1880.			
February 14..	Amount from state treasurer...............	1,415 79	
July 31........	Amount from state treasurer...............	1,852 93	
September 30.	Balance, overdraft..........................	129 53	

Cr.

1879.			
October 1.....	By balance, overdraft.......................		$2,022 45
1880.			
September 30.	Superintendent's orders paid		3,958 24
		$5,980 69	$5,980 69
1880.			
October 1.....	By balance.................................		$129 53

IMPROVEMENT OF GROUNDS.
Dr.

1879.			
October 31....	To amount from state treasurer............	$108 50	
1880.			
February 14..	Amount from state treasurer...............	751 65	
July 31........	Amount from state treasurer...............	74 54	
September 30.	Balance, overdraft..........................	308 04	

Cr.

1879.			
October 1.....	By balance, overdraft.......................		$108 50
1880.			
September 30.	Superintendent's orders paid		1,134 23
		$1,242 73	$1,242 73
1880.			
October 1.....	By balance.................................		$308 04

Treasurer's Report—*Continued.*

NEW KITCHEN.

Dr.

1879.		
October 31....	To amount from state treasurer........................	$1,244 58
1880.		
February 2...	Amount from state treasurer.......................	745 87
April 23........	Amount from state treasurer.......................	151 60
July 31.......	Amount from state treasurer.......................	838 00

Cr.

1879.			
October 1.....	By balance, overdraft................................		$357 03
1880.			
September 30.	Superintendent's orders paid.....................		2,623 02
		$2,980 05	$2,980 05

WATER SUPPLY.

Dr.

1879.		
October 31....	To amount from state treasurer....................	$1,020 68
1880.		
February 2...	Amount from state treasurer.......................	152 54
April 23.......	Amount from state treasurer.......................	180 25
September 30.	Balance, overdraft................................	77 25

Cr.

1879.			
October 1.....	By balance, overdraft..............................		$1,020 68
1880.			
September 30.	Superintendent's orders paid.....................		410 04
		$1,430 72	$1,430 72
1880.			
October 1.....	By balance..		$77 25

EXTENDING SEWER.

Dr.

1879.		
October 31....	To amount from state treasurer....................	$1,133 84
1880.		
February 2...	Amount from state treasurer.......................	101 14
April 23.......	Amount from state treasurer.......................	17 04
September 30.	Balance, overdraft................................	11 25

Cr.

1879.			
October 1.....	By balance, overdraft.............................		$921 22
September 30.	Superintendent's orders paid.....................		342 05
		$1,263 27	$1,263 27
1880.			
October 1.....	By balance..		$11 25

REMOVAL OF BARN.

Dr.

1879.		
October 31....	To amount from state treasurer....................	$511 00
1880.		
February 2...	Amount from state treasurer.......................	413 70
April 23.......	Amount from state treasurer.......................	9 00
July 31........	Amount from state treasurer.......................	66 30

Cr.

1879.			
October 1.....	By balance, overdraft................................		$175 00
1880.			
September 30.	Superintendent's orders paid		825 00
		$1,000 00	$1,000 00

Treasurer's Report—*Continued.*

	RECAPITULATION.		
	Dr.		
1880. October 1.....	To balance, ordinary expense...........................	$13,567 30	
	Cr.		
1880. September 80.	By overdraft, improvement and repairs...............	....	$129 53
	Overdraft, improvement of grounds...............	....	308 04
	Overdraft, water supply....................	....	77 25
	Overdraft, extending sewer....................	....	11 25
	Total balances on hand....................	....	13,041 23
		$13,567 30	$13,567 30

INVENTORY.

Showing the Value of all Property belonging to the Southern Hospital for the Insane, at Anna, September 30, 1880.

Food supplies	$3,591 37
Clothing, bedding, dry goods, etc	8,076 63
Laundry apparatus and supplies	496 30
Fuel	367 50
Light, supplies for	167 36
Medicines and medical supplies	814 19
Books and stationery	849 94
Printing, blank forms, etc	90 50
Musical instruments, fixtures, etc	2,243 55
Instruments and apparatus	526 05
Household apparatus and supplies	1,105 03
Furniture	15,494 50
Buildings, land and building material	477,306 39
Tools	320 75
Machinery, etc	16,151 40
Farm and garden implements, stock, etc	5,009 71
Shop supplies	12 17
Total	**$532,653 34**

LAW OF ADMISSION.

CHAPTER 85, REVISED STATUTES 1874, PAGE 681, ENTITLED "LUNATICS."

AN ACT to revise the law in relation to the commitment and detention of Lunatics, [Approved March 24, 1874. In force July 1, 1874.]

PETITION.] § 1. *Be it enacted by the people of the state of Illinois, represented in the general assembly,* That when any person is supposed to be insane or distracted, any near relative, or in case there be none, any respectable person residing in the county, may petition the judge of the county court for proceedings to inquire into such alleged insanity or distraction. For the hearing of such application, and proceedings thereon, the county court shall be considered as always open.

WRIT—SERVICE.] § 2. Upon the filing of such petition, the judge shall order the clerk of the court to issue a writ, directed to the sheriff, or any constable, or the person having the custody or charge of the alleged insane or distracted person, unless he shall be brought before the court without such writ, requiring the alleged insane person to be brought before him at a time and place to be appointed for the hearing of the matter. It shall be the duty of the officer or person to whom the writ is directed, to execute and return the same, and bring the alleged insane person before the court as directed in the writ.

SUBPENAS.] § 3. The clerk shall also issue subpenas for such witnesses as may be desired on behalf of the petitioner, or of the person alleged to be insane, to appear at the time fixed for the trial of the matter.

JURY—TRIAL.] § 4. At the time fixed for the trial, a jury of six persons, one of whom shall be a physician, shall be impaneled to try the case. The case shall be tried in the presence of the person alleged to be insane, who shall have the right to be assisted by counsel, and may challenge jurors as in civil cases. The court may, for good cause, continue the case from time to time.

VERDICT—FORM.] § 5. After hearing the evidence, the jury shall render their verdict in writing, signed by them, which shall embody the substantial facts shown by the evidence, which verdict may be substantially in the following form:

63

STATE OF ILLINOIS, } ss.
.... County. }

We, the undersigned, jurors in the case of.............(naming the person alleged to be insane), having heard the evidence in the case, are satisfied that said ....is insane, and is a fit person to be sent to a state hospital for the insane; that he is a resident of the state of Illinois, and county of................; that his age is..........; that his disease is of..........duration; that the cause is supposed to be.............(or is unknown); that the disease is (or is not) with him hereditary; that he is not (or is) subject to epilepsy, and that he does (or does not) manifest homicidal or suicidal tendencies. (If the person be a pauper, the fact shall also be announced in the verdict.)

VERDICT RECORDED—ORDER OF COMMITTAL—APPLICATION.] § 6. Upon the return of the verdict, the same shall be recorded at large by the clerk, and if it appears that the person is insane, and is a fit person to be sent to a state hospital for the insane, the court shall enter an order that the insane person be committed to a state hospital for the insane, and thereupon it shall be the duty of the clerk of the court to make application to the superintendent of some one of the state hospitals for the insane, for the admission of such insane person.

TO WHICH HOSPITAL—APPLICATION, ETC.] § 7. If such insane person is a pauper, the application shall be first made to the nearest hospital, but if he be not a pauper, application shall be made to such one of the state hospitals for the insane as the relatives or friends of the patient shall desire. In any case, if, on account of the crowded condition of any one of the hospitals, or for other good reason, the patient cannot be received therein, or it is not desirable to commit him thereto, he may be committed to any other of said hospitals. Upon receiving any such application, the superintendent shall immediately inform the clerk whether the patient can be received, and if so, at what time; and if not, shall state the reason why.

WARRANT TO COMMIT.] § 8. Upon receiving notice at what time the patient will be received, the clerk shall, in due season for the conveyance of the person to the hospital by the appointed time, issue a warrant, directed to the sheriff or any other suitable person, preferring some relative of the insane person when desired, commanding him to arrest such insane person and convey him to the hospital; and if the clerk is satisfied that it is necessary, he may authorize an assistant to be employed.

FORM OF WARRANT.] § 9. The warrant may be substantially as follows:

STATE OF ILLINOIS, } ss.
....................County. }

The People of the State of Illinois: to.............................

You are hereby commanded forthwith to arrest ...., who has been declared to be insane, and convey him to the Northern (or as the case may be) Illinois Hospital for the Insane (and you are hereby authorized to take to your aid an assistant, if deemed necessary), and of this warrant make due return to his office after its execution.

Witness my hand and the seal of the county court of .... county, this .... day of ...., A. D. ......

[L. S.]   Clerk of the County Court ....County.

INDORSEMENT—RETURN.] § 10. Upon receiving the patient, the superintendent shall indorse upon said warrant a receipt, as follows:

Northern (or as the case may be) Illinois Hospital for the Insane.
Received this ...... day of ...., A. D. ......, the patient named in the within warrant.
....
Superintendent

This warrant, with a receipt thereon, shall be returned to the clerk, to be filed by him with the other papers relating to the case.

WHO NOT ADMITTED—IDIOTS DISCHARGED.] § 11. No person having any contagious or infectious disease, and no idiot, shall be admitted to either of the state hospitals. When the trustees and superintendent shall find that an idiot has been received into the hospital, they may discharge him.

TEMPORARY COMMITMENT.] § 12. If the court shall deem it necessary, pending proceedings and previous to verdict, or after verdict and pending admission to the hospital, temporarily to restrain of his liberty the person alleged to be insane, then the court shall make such order in that behalf as the case may require, and the same being entered of record, a copy thereof, certified by the clerk, shall authorize such person to be temporarily detained by the sheriff, jailer or other suitable person to whom the same shall be directed.

COSTS.] § 13. When a person not a pauper is alleged to be insane, and is found by the jury not to be insane, the costs of the proceeding, including the fees of the jury, shall be paid by the petitioner, and judgment may be awarded against him therefor. If such person is found to be insane, such costs shall be paid by his guardian, conservator or relatives, as the court may direct. If the person alleged to be insane is a pauper, the costs of the proceeding, including the fees of the jury, shall be paid out of the county treasury: *Provided*, if such pauper is found not to be insane, the court may, in its discretion, award the costs against the petitioner.

WHO TO PAY EXPENSES—SHERIFF'S FEES.] § 14. The expense of conveying a pauper to the hospital shall be paid by the county in which he resides, and that of any other patient by his guardian, conservator or relatives; and in no case shall any such expense be paid by the state, or out of any funds for the insane. The fees of the sheriff for conveying any person to a hospital shall be the same as for conveying convicts to the penitentiary.

BOND TO FURNISH CLOTHING.] § 15. If the person be not a pauper, then one or more persons, relatives or friends of the patient, shall, upon his admission into the hospital, become responsible to the trustees for finding the patient in clothes, and removing him when required; and shall execute a bond conditioned as follows, viz:

Know all men by these presents, that we.......and.......of the county of.......and state of Illinois, are held and firmly bound unto the trustees of the Northern (or as the case may be) Illinois Hospital for the Insane in the sum of one hundred dollars ($100), for the payment of which we jointly and severally bind ourselves firmly by these presents.

The condition of this obligation is, that whereas........, insane person, of the county and state aforesaid, has been admitted as a patient into said Hospital for the Insane: Now, therefore, if we shall find said patient in suitable and sufficient clothing whilst ...may remain in said institution, and shall promptly pay for such articles of clothing as it may be necessary to procure for said........at the hospital, and shall remove.... from said hospital when required by the trustees to do so, then this obligation to be void; otherwise to remain in full force.

Witness our hands and seals, this....day of....A. D....

.... [SEAL]
.... [SEAL]

CLOTHING.] § 16. The clothing to be furnished each patient, upon being sent to the hospital, shall not be less than the following: For a male, three new shirts, a new and substantial coat, vest, and two pairs of pantaloons of woolen cloth, three pairs of woolen socks, a black or dark stock or cravat, a good hat or cap, and a pair of new shoes or boots, and a pair of slippers to wear within doors. For a female, in addition to the same quantity of undergarments, shoes and stockings, there shall be two woolen petticoats or skirts, three good dresses, a cloak or shawl, and a decent bonnet. Unless such clothing be delivered, in good order, to the superintendent, he shall not be bound to receive a patient.

PAUPERS—COUNTY TO FURNISH CLOTHING, ETC.] § 17. If the insane person be a pauper, it shall be the duty of the judge of the county court to see that he is furnished with the necessary amount of substantial clothing at the time he is sent to the hospital, and from time to time while he remains a patient in the hospital, and that he be removed therefrom when required by the trustees; the expense of such clothing and removal shall be paid out of the county treasury, upon the certificate of the judge of the county court.

DISCHARGE OF PATIENT—NOTICE—REMOVAL.] § 18. Whenever the trustees shall order any patient discharged, the superintendent shall at once notify the clerk of the county court of the proper county thereof, if the patient is a pauper, and if not, shall notify all the persons who signed the bond required in section 15 of this act, and request the removal of the patient. If such patient be not removed within thirty days after such notice is received, then the superintendent may return him to the place from whence he came, and the reasonable expenses thereof may be recovered by suit on the bond, or in case of a pauper, shall be paid by the proper county.

NON-RESIDENT PATIENTS.] § 19. Whenever application shall be made for a patient not residing within the state, if the superintendent shall be of opinion that from the character of the case it is probably curable, and if there be at the time any room in the hospital, the trustees, in their discretion, may order the patient to be admitted, always taking a satisfactory bond for the maintenance of the patient, and for his removal when required. The rate of maintenance in such cases shall be fixed by the trustees, and two months' pay in advance shall be required. But no such patient shall be detained without the order of a court of competent jurisdiction, or a verdict of a jury.

RESTORATION TO REASON—DISCHARGE.] § 20. When any patient shall be restored to reason, he shall have the right to leave the hospital at any time, and if detained therein contrary to his wishes after such restoration, shall have the privilege of a writ of *habeas corpus* at all times, either on his own application, or that of any other person in his behalf. If the patient is discharged on such writ, and if it shall appear that the superintendent has acted in bad faith, or negligently, the superintendent shall pay all the cost of the proceeding. Such superintendent shall moreover be liable to a civil action for false imprisonment.

COUNTY HOSPITAL.] § 21. This act shall not be construed to prevent the committing of any insane pauper to the hospital for the insane of the county in which he may reside, where such a hospital is provided.

TRIAL BY JURY NECESSARY.] § 22. No superintendent, or other officer or person connected with either of the state hospitals for the insane, or with any hospital or asylum for insane or distracted per-
—5

sons in this state, shall receive, detain or keep in custody, at such hospital or asylum, any person who shall not have been declared insane by the verdict of a jury, and authorized to be confined by the order of a court of competent jurisdiction; and no trial shall be had of the question of the sanity or insanity of any person before any judge or court, without the presence of the person alleged to be insane.

PENALTY.] § 23. If any superintendent, or other officer or person connected with either of the state hospitals for the insane, or with any hospital or asylum for insane or distracted persons, in this state, whether public or private, shall receive or detain any person who has not been declared insane by the verdict of a jury, and whose confinement is not authorized by the order of a court of competent jurisdiction, he shall be confined in the county jail not exceeding one year, or fined not exceeding $500, or both, and be liable civilly to the person injured, for all the damages which he may have sustained; and if he be connected with either of the insane hospitals of this state, he shall be discharged from service therein.

www.ingramcontent.com/pod-product-compliance
Lightning Source LLC
Chambersburg PA
CBHW022006190326
41519CB00010B/1398